ONE WEEK

PRACTICAL NETWORK DESIGN TECHNIQUES

Second Edition

PRACTICAL NETWORK DESIGN TECHNIQUES

Second Edition

A Complete Guide for WANs and LANs

Gilbert Held
S. Ravi Jagannathan

AUERBACH PUBLICATIONS

A CRC Press Company
Boca Raton   London   New York   Washington, D.C.

Library of Congress Cataloging-in-Publication Data

Held, Gilbert, 1943-
    Practical network design techniques / Gilbert Held, S. Ravi Jagannathan.
      p. cm.
    Includes index.
    ISBN 0-8493-2019-4 (alk. paper)
      1. Data transmission systems. 2. Computer networks. I. Jagannathan, S. Ravi, II. Title.

TK5105.H4285 2004
004.6'6--dc22

2004041146

Visit the Auerbach Web site at www.auerbach-publications.com

© 2004 by CRC Press LLC
Auerbach is an imprint of CRC Press LLC

No claim to original U.S. Government works
International Standard Book Number 0-8493-2019-4
Library of Congress Card Number 2004041146
Printed in the United States of America 1 2 3 4 5 6 7 8 9 0
Printed on acid-free paper

CONTENTS

1. Introduction 1
  1.1 Rationale 2
  1.2 Book Overview 2
       1.2.1 WAN Design 2
       1.2.2 LAN Design 4

SECTION I: WIDE AREA NETWORK DESIGN TECHNIQUES

2. Transmission Characteristics 9
  2.1 Communications Constraints 9
       2.1.1 Throughput 9
       2.1.2 Response Time 10
  2.2 Information Transfer Rate 11
       2.2.1 Delay Time 12
       2.2.2 Other Delay Factors 12
  2.3 BISYNC Communications Protocol Models 13
       2.3.1 Error Control 14
       2.3.2 Half-Duplex Throughput Model 15
              2.3.2.1 Computation Examples 17
              2.3.2.2 Block Size versus Error Rate 20
       2.3.3 Return Channel Model 23
       2.3.4 Full-Duplex Model 24

3. Wide Area Network Line Facilities and Economic Trade-Offs 27
  3.1 Line Connections 27
       3.1.1 Dedicated Line 28
       3.1.2 Leased Line 28
       3.1.3 Switched Line 31
  3.2 Leased versus Switched Facility Analysis 32
       3.2.1 Servicing Additional Terminal Devices 33
       3.2.2 Line Sharing Economics 34
       3.2.3 Concepts to Consider 40

3.3   Additional Analog Facilities to Consider 43
 3.3.1   WATS 43
 3.3.2   Foreign Exchange 48
  3.3.2.1   Economics of Use 49
3.4   Digital Facilities 51
 3.4.1   AT&T Offerings 51
 3.4.2   European Offerings 52
  3.4.2.1   Data Service Units 52
 3.4.3   Economics of Use 54
 3.4.4   Dataphone Digital Service 54
 3.4.5   Accunet T1.5 56
  3.4.5.1   DDS as Opposed to Accunet T1.5 57
 3.4.6   Accunet Spectrum of Digital Services 57
  3.4.6.1   Comparison to Other Digital Services 58
  3.4.6.2   Accunet T45 61

4.   Multiplexing, Routing, and Line Sharing Equipment 63
4.1   Multiplexers and Data Concentrators 63
 4.1.1   Evolution 64
 4.1.2   Comparison with Other Devices 64
 4.1.3.   Device Support 66
 4.1.4   Multiplexing Techniques 66
  4.1.4.1   Frequency Division Multiplexing 66
  4.1.4.2   Time Division Multiplexing 68
 4.1.5   The Multiplexing Interval 70
 4.1.6   TDM Techniques 70
 4.1.7   TDM Constraints 72
 4.1.8   TDM Applications 76
  4.1.8.1   Series Multipoint Multiplexing 77
  4.1.8.2   Hub-Bypass Multiplexing 78
  4.1.8.3   Front-End Substitution 79
  4.1.8.4   Inverse Multiplexing 80
 4.1.9   Multiplexing Economics 81
 4.1.10   Statistical and Intelligent Multiplexers 83
  4.1.10.1   Statistical Frame Construction 84
  4.1.10.2   Buffer Control 87
  4.1.10.3   Service Ratio 87
  4.1.10.4   Data Source Support 89
  4.1.10.5   Switching and Port Contention 90
  4.1.10.6   Intelligent Time Division Multiplexes 91
  4.1.10.7   STDM/ITDM Statistics 91
  4.1.10.8   Using System Reports 92
  4.1.10.9   Features to Consider 96
4.2   Routers 97

4.2.1   Functionality 97
4.2.2   Ports and Connectors 98
4.2.3   Address Resolution 98
4.2.4   Hardware and Software 99
        4.2.4.1   Software Modules 100
4.2.5   Data Flow and Packet Switching 100
4.3 Modem and Line Sharing Units 102
4.3.1   Operation 103
4.3.2   Device Differences 104
4.3.3   Sharing Unit Constraints 105
4.3.4   Summary 107

5. Locating Data Concentration Equipment 109
5.1 Graph Theory and Network Design 109
5.1.1   Links and Nodes 110
5.1.2   Graph Properties 111
        5.1.2.1   Network Subdivisions 112
        5.1.2.2   Routes 112
        5.1.2.3   Cycles and Trees 113
5.1.3   The Basic Connection Matrix 113
        5.1.3.1   Considering Graph Weights 113
5.2 Equipment Location Techniques 115
5.2.1   Examining Distributed Terminals 116
5.2.2   Using a Weighted Connection Matrix 117
5.2.3   Automating the Location Process 118
5.2.4   Extending the Node Location Problem 120
5.2.5   Switched Network Utilization 120
5.2.6   Other Program Modifications 123

6. Multidrop Line Routing Techniques 127
6.1 Multidrop Routing Algorithms 127
6.1.1   The MST Technique 128
        6.1.1.1   Applying Prim's Algorithm 129
        6.1.1.2   Considering Fan-Out 132
6.1.2   Modified MST Technique 132
6.1.3   Using a Connection Matrix 133
6.2 Automating the MST Process 135
6.2.1   Basic Language Program 136
6.3 Considering Network Constraints 140
6.3.1   Terminal Response Time Factors 140
6.3.2   Estimating Response Time 141
6.3.3   Front-End Processing Limitations 142
6.4 Summary 142

7. Sizing Communications Equipment and Line Facilities. 143
7.1 Sizing Methods. 144
     7.1.1 Experimental Modeling 144
     7.1.2 The Scientific Approach. 146
7.2 Telephone Terminology Relationships. 146
     7.2.1 Telephone Network Structure. 147
          7.2.1.1 Trunks and Dimensioning. 147
          7.2.1.2 The Decision Model 149
7.3 Traffic Measurements 150
     7.3.1 The Busy Hour 150
     7.3.2 Erlangs and Call-Seconds. 151
     7.3.3 Grade of Service 154
     7.3.4 Route Dimensioning Parameters. 155
     7.3.5 Traffic Dimensioning Formulas. 156
7.4 The Erlang Traffic Formula. 156
     7.4.1 Computing Lost Traffic. 159
     7.4.2 Traffic Analysis Program 161
     7.4.3 Traffic Capacity Planning. 165
     7.4.4 Traffic Tables. 171
          7.4.4.1 Access Controller Sizing 173
7.5 The Poisson Formula 177
     7.5.1 Access Controller Sizing. 178
     7.5.2 Formula Comparison and Utilization 182
     7.5.3 Economic Constraints. 183
7.6 Applying the Equipment Sizing Process 184

SECTION II: LOCAL AREA NETWORK DESIGN TECHNIQUES

8. Local Area Network Devices. 189
8.1 Stations and Segment. 189
8.2 Repeaters. 190
8.3 Hubs 190
8.4 Bridges 191
     8.4.1 Types of Bridges. 191
     8.4.2 The Learning Bridge 192
     8.4.3 Translational Bridges 195
     8.4.4 Bridged Topologies 195
     8.4.5 SR Bridge 196
     8.4.6 SRT Bridges. 196
     8.4.7 SR/TL Bridges 196
     8.4.8 Remote Bridges 197

8.5   LAN Switches................................... 197
      8.5.1   The Basic Premise......................... 197
      8.5.2   Delay Timing.............................. 198
      8.5.3   Types of LAN Switches 198
      8.5.4   Switch Design Methods 199
      8.5.5   Network Utilization 200
8.6   Routers 200
      8.6.1   Routers in Relation to LANs............... 201
      8.6.2   Router Behavior........................... 201
      8.6.3   Types of Routers 202
      8.6.4   Route Protocols 203
      8.6.5   Routing Protocols 204
              8.6.5.1   Vector Distance................. 204
              8.6.5.2   Link State 205
8.7   Brouters...................................... 205
8.8   Gateways...................................... 205
8.9   Network Interface Cards....................... 206
8.10 File Servers 206

9. Local Area Network Topologies 209
9.1   Introduction.................................. 209
9.2   Key Topologies 213
      9.2.1   The Loop Topology......................... 213
      9.2.2   The Bus Topology.......................... 213
      9.2.3   The Tree Topology 214
      9.2.4   Star Topology 214
      9.2.5   High Data Rates and the Associated Problems 216
9.3   Pros and Cons of Different Topologies.......... 217
      9.3.1   Loop 218
      9.3.2   Bus or Tree............................... 218
      9.3.3   Star 219
9.4   Structured Cabling System 219
9.5   MAC Protocols................................. 220
      9.5.1   Carrier Sense Multiple Access with Collision
              Detection................................. 220
      9.5.2   Carrier Sense Multiple Access with Collision
              Avoidance 222
      9.5.3   Token Passing............................. 222
      9.5.4   Switched, Connection-Oriented MAC 222
      9.5.5   Demand Priority Media Access............... 222
9.6   LAN Architecture Evolution 223
      9.6.1   Architectural Design of LANs.............. 224
              9.6.1.1   Bottom-Up 225
              9.6.1.2   Top-Down 225

9.7 Geometric Limitations of LANs 225
9.8 Fiber Channel Topologies. 226

**10. A Tutorial on the Ethernet Family of Local
Area Networks** 229
   10.1 Introduction 229
   10.2 Transmission Media 230
      10.2.1 Twisted Pair 230
      10.2.2 Coaxial Cable 231
         10.2.2.1 Coaxial Adapters. 232
      10.2.3 Optical Fiber Cable 232
         10.2.3.1 Fiber Optic Technology 233
   10.3 An Excursion into the Ethernet Family 234
      10.3.1 10 Mbps LANs. 234
         10.3.1.1 10BASE-5 235
         10.3.1.2 10BASE-2 235
         10.3.1.3 10BASE-T 235
         10.3.1.4 10BROAD-36 236
         10.3.1.5 10BASE-F 236
      10.3.2 Fast Ethernet (100 Mbps). 236
         10.3.2.1 Backbone Operation. 237
         10.3.2.2 Switch Segmentation. 238
      10.3.3 Gigabit Ethernet (1000 Mbps) 238
      10.3.4 10 Gigabit Ethernet 241
   10.4 LAN Ethernet Design 243
      10.4.1 Campuswide VLANs with Multilayer Switching. . . . 245
   10.5 Switches Revisited 246
      10.5.1 Scalability, Latency, Global Effect of Failures
             and Collisions 247
      10.5.2 Encoding Schemes. 248
         10.5.2.1 Nonreturn to Zero Level 248
         10.5.2.2 Nonreturn to Zero Invert on 1s. 249
         10.5.2.3 Manchester. 249
         10.5.2.4 Differential Manchester 249
         10.5.2.5 4B/5B-NRZ-I 249
         10.5.2.6 MLT-3. 250
         10.5.2.7 8B/10B 251

11. Ethernet Performance Characteristics 253
   11.1 Introduction. 253
   11.2 Frame Operations. 253
      11.2.1 Ethernet Frames. 254
         11.2.1.1 Preamble 255
         11.2.1.2 SOF Delimiter. 255

11.2.1.3 Source and Destination Addresses 255
11.2.1.4 Type 256
11.2.1.5 Length 256
11.2.1.6 Data Field 256
11.2.1.7 Frame Check Sequence 257
11.2.2 Fast Ethernet Frames 257
11.2.3 Gigabit Ethernet Frames 257
11.2.3.1 Carrier Extension 258
11.2.3.2 Packet Bursting 258
11.2.4 Frame Overhead 259
11.3 Availability Levels 259
11.4 Network Traffic Estimation 261
11.5 An Excursion into Queuing Theory 263
11.5.1 Buffer Memory Considerations 264
11.6 Ethernet Performance Details 266
11.6.1 Network Frame Rate 266
11.6.2 Gigabit Ethernet Considerations 267
11.6.3 Actual Operating Rate 268
11.7 Bridging a Network 268

12. Issues at the Network, Transport, and
    Application Layers 271
12.1 Internetworking Overview 271
12.2 Protocol Architecture 273
12.3 Design Issues 274
12.3.1 Addressing 274
12.3.2 Routing 275
12.3.3 Datagram Lifetime 275
12.3.4 Fragmentation or Reassembly 275
12.4 Routing and Route Protocols 276
12.5 Routing Revisited 277
12.5.1 Routing Protocols 279
12.5.2 DV Protocols 280
12.5.3 LS Protocols 281
12.6 Excursion into the Transport Layer 283
12.7 Multimedia Service 284
12.8 Some Delay Time Calculations 285
12.8.1 10 Mbps Ethernet, 100 Mbps Fast Ethernet,
       and 1000 Mbps Gigabit Ethernet 285
12.8.2 Switches 286

13. Wireless Local Area Networks 287
13.1 Introduction 287
13.2 Media Considerations 288

13.2.1 IR Systems. 288
 13.2.1.1 Directed Beam IR. 288
 13.2.1.2 Omnidirectional IR 288
 13.2.1.3 Diffused IR. 289
13.2.2 RF LAN Networks 289
 13.2.2.1 Spread Spectrum. 289
 13.2.2.2 Spread Spectrum Configuration. 291
 13.2.2.3 Narrowband RF 291
13.3 Transmission Issues 292
13.4 WLAN Topology. 292
13.5 Wireless Standards 294
13.5.1 Independent Configuration. 295
13.5.2 Infrastructure Configuration 296
13.6 WLAN Design Considerations. 296
13.7 Wireless LAN Switching 297
13.7.1 Additional Functions of WLAN Switches. 299

14. Local Area Network Internetworking Issues 301
14.1 Introduction. 301
14.2 Overview of Internetworking Concepts. 301
14.3 Switching Overview 302
14.4 The Tiered (Layered) Approach 304
14.5 Evaluating Backbone Capabilities 305
14.5.1 Path Optimization 305
14.5.2 Traffic Prioritizing 306
 14.5.2.1 Priority Queuing. 306
 14.5.2.2 Custom Queuing. 306
 14.5.2.3 Weighted Fair Queuing. 308
14.5.3 Load Splitting 308
14.5.4 Alternative Paths 309
14.5.5 Encapsulation (Tunneling) 309
14.6 Distribution Services. 309
14.6.1 Backbone Bandwidth Management 309
14.6.2 Area and Service Filtering 310
14.6.3 Policy-Based Distribution. 310
14.6.4 Interprotocol Route Redistribution 310
14.6.5 Media Translation 311
14.7 Local Access Services. 311
14.7.1 Value-Added Addressing 312
14.7.2 Network Segmentation. 312
14.7.3 Broadcast or Multicast Capabilities. 312
14.7.4 Naming, Proxy, and Local Cache Capabilities. 313
14.7.5 Media Access Security 313
14.7.6 Router Discovery. 314

14.7.7 Internet Control Message Protocol Router
Discovery 314

14.7.8 Proxy ARP. 314

14.7.9 Routing Information Protocol. 314

14.8 Constructing Internets by Design 314

14.9 Using Switches (Revisited) 315

14.9.1 Comparison of Switches and Routers 315

Index 317

PREFACE

One of the numerous tasks associated with managing a nationwide data communications network is the training of employees. Although there are many fine schools and seminar companies that teach a variety of data communications related topics, conspicuous by their absence are courses or seminars that provide a practical guide to the design of wide area networks (WANs) and local area networks (LANs). Recognizing this void, Gil Held originally developed a series of short monographs and lecture notes that he used as a basis to disseminate information to readers of a communications-oriented trade publication in the form of articles, as well as conducting informal lectures for employees covering a variety of network design techniques. This initial activity occurred during the early 1980s, prior to the active deployment of LANs and resulted in the first edition of *Practical Network Design Techniques,* which was published in 1991 and reprinted in 1994 and 1996.

Although focused upon the WAN, the first edition of *Practical Network Design Techniques* included some information that was applicable to the design of LANs and to a degree indicated the close relationship between the two dissimilar networks with respect to the design process. For example, in a WAN when multiplexers and concentrators were used to support dial-in activity from remote users, the sizing of lines, modems, and multiplexer ports occurred using Erlang and Poisson traffic formulas. In comparison, a similar problem in a LAN environment occurred when organizations needed to support remote dial-in onto their LAN, to include the use of data concentrators by Internet service providers (ISPs) for supporting customers accessing the Internet.

This new edition of *Practical Network Design Techniques* is subdivided into two sections — one focused on the WAN, while the other is oriented to LANs. Because the LAN war was in effect won by Ethernet, the second section of this book is focused upon Ethernet networks. Because some areas of network design are applicable to both wide and local area networks, considerable thought was required concerning where to place some portions of the original edition of this book in the new edition. After considerable thought, we decided to incorporate the bulk of the initial edition of this book into the first section of this new edition while indicating in revisions to the text the manner by which certain information can also be used in the LAN design process.

In addition to a new section covering LAN, this book has a new coauthor, S. Ravi Jagannathan. Thus, readers can now gain from the background and experience of two authors as they read this new edition.

College students can use this book as a text to obtain an understanding of the major type of network design problems and their solutions as well as a reference for the data communications practitioner. To maximize the use of this book, we included a number of practical networking problems and their solutions, as well as examples of methods to perform economic comparisons between different communications services and hardware configurations. Because we are quite concerned about reader feedback, we encourage you to send your comments concerning the contents of this book. Let us know if our coverage of material was appropriate, if some areas require additional coverage, or if we failed to cover an area of interest. You can contact us either through our publisher whose Web address is on the back cover of this book or via e-mail at the addresses listed below.

Gilbert Held
Macon, Georgia
gil_held@yahoo.com

S. Ravi Jagannathan
NSW, Australia
jagapink@yahoo.co.uk

ACKNOWLEDGMENTS

The development of a book in many ways is similar in scope to the winning season of a sports team — success depends upon the effort of many individuals. The publication of this new edition is no exception. Thus, Gil Held would like to thank his publisher, Rich O'Hanley of CRC Press, for backing this new edition as well as for coordinating the numerous steps required to convert a manuscript into the book you are reading. Concerning the manuscript, because the original disk files from the first edition aged too well over the years, the original portion of the first edition of this book covering WAN design had to be retyped. Fortunately, Gil's wife Beverly "volunteered" her typing skills for a good cause. Ravi acknowledges the support and encouragement of his wife Dee in all aspects of the production of this edition.

1

INTRODUCTION

Network design is both a science and an art. The scientific foundation for network design is based upon mathematics that provide the applicable tools for solving such problems as:

- Determining the circuit operating rate required to interconnect two geographically separated LANs
- Deciding where to locate data concentration equipment
- Routing a multidrop line to minimize the cost associated with a circuit configured to connect multiple network locations onto a common transmission facility

The art associated with network design involves one having knowledge of the capabilities of different types of LAN and WAN communications equipment and how such capabilities can be used to develop a variety of network configurations.

In this new edition of *Practical Network Design Techniques,* we will focus the majority of our attention upon the scientific aspect of network design, using mathematical models to illustrate important networking concepts, as well as to solve many common network design problems. Although we will also examine the capabilities and limitations of several types of data communications equipment, our coverage of the art side to network design will not be as comprehensive as our examination of the scientific aspect to network design.

This new edition expands upon the original edition of this book, adding a new section focused upon LANs. Because the interconnection of geographically separated LANs via a WAN depends upon many LAN and WAN factors, the decision where to place such information required considerable thought. Thus, this introductory chapter serves as a guide to the contents of this book, as well as explains the reason for the placement of some topics whose location in a particular section required considerable thought.

1.1 RATIONALE

Although the title of this book references network design, its rationale includes more than simply focusing on that topic. We will examine cost and performance issues that, although falling under the network design umbrella, can also be used by network analysts and designers as tools and techniques to follow to optimize existing networks with respect to their cost or performance. In several instances, we will examine reports generated by communications equipment to obtain an understanding of how different report elements can be used to tailor the cost or performance of a network segment. Although the large disparity between the uses of different types of communications equipment in different networks may result in the examination of report parameters not being applicable to some readers, the concept concerning the use of equipment reports should be applicable to all readers.

1.2 BOOK OVERVIEW

As previously mentioned, this book is subdivided into two sections: the first section primarily focuses on WAN design techniques and the second section is oriented toward LAN design techniques. For continuity purposes, chapters are numbered sequentially throughout this book.

Please note the distinction between "internet" (lowercase i) and "Internet" (capital I) used throughout this book. An "internet" is a connection of subnetworks, whereas, the "Internet" is a global network connecting millions of users, commonly known as the World Wide Web.

To obtain an appreciation of the scope and depth of the contents of this book we will briefly tour succeeding chapters of this book. Although each chapter was written to be as independent as possible of previous and succeeding chapters, readers should note that in some instances material in one chapter was required to reference material in another chapter. Even though chapter independence allows readers to turn to material of specific interest, for most readers and especially those entering the network design profession it is recommended that you read the chapters in this book in their sequence of presentation. That said, grab a favorite drink and perhaps a few munchies and follow us into the wonderful world of network design.

1.2.1 WAN Design

In Chapter 2, we will focus our attention upon the transmission characteristics of batch and interactive transmissions systems. After examining the characteristics of each system, we will develop several models that can be used to predict performance, as well as to provide readers with a foundation for understanding key communications concepts.

To obtain a firm understanding of the economics associated with different types of communications facilities, we will examine the cost associated with the use of several types of analog and digital circuits and switched network facilities in Chapter 3. In this chapter, we will use tariffs that were in effect when this new book edition was developed. Although the use of those tariffs illustrates several economic trade-off concepts, readers are cautioned to obtain the latest tariff for each line facility under consideration to ensure the validity of their economic analysis.

Because three of the most commonly utilized types of WAN communications equipment are multiplexers, routers, and line sharing devices, we will examine their use in Chapter 4. In this chapter, we will examine the economics associated with the use of the previously mentioned equipment, as well as their operational characteristics and constraints associated with their use.

In Chapter 5, we will examine the use of graph theory to solve equipment location problems, such as determining the location to install a remote multiplexer or router based upon the geographical distribution of LANs or terminal devices required to be supported. Regardless of whether your organization uses multiplexers or routers to service a series of remote locations, a common problem is their placement to minimize the cost of communications facilities.

Because digital circuits long ago replaced the use of analog circuits for the backbone of most communications carriers, one might expect that there is no need to consider multidrop lines. Although most organizations long ago converted to digital circuits, to paraphrase Mark Twain, the death of analog has been greatly exaggerated. Cable modems and digital subscriber lines (DSL) represent analog technology and in many locations, the proverbial last mile continues to use analog technology. Because the routing of the backbone common circuit to support many last mile solutions represents a multidrop line, we will continue our use of graph theory in Chapter 6, examining the application of this area of mathematics to developing an optimally routed multidrop line.

We will conclude our coverage of WAN design by focusing on an area that also applies to an area of LAN design. In Chapter 7, we will use another area of mathematics originally developed for telephone traffic engineering to determine the appropriate number of ports for different types of data concentration equipment. The process covered in this chapter is more formally known as equipment sizing and can also be applied to determine the number of lines to install at a rotary or hunt group. Because dial-in lines can be connected to a variety of equipment to include routers, multiplexers, and data concentrators, the equipment sizing process covered in this chapter can assist organizations in determining the level of support required for certain types of WAN and LAN design efforts.

1.2.2 LAN Design

In this section of *Practical Network Design Techniques,* we turn our attention to LANs. Beginning with coverage of the definition of a LAN and the rationale for its use, we will then become acquainted with the members of the Ethernet family, which are the mainstay of Section 2. This is followed by an examination of the design constraints and performance considerations associated with data flow at Layer 2, Layer 3, and above. In doing so, we implicitly cover the key issues, problems, and trends that arise when we attempt to design a LAN solution for any given local networking scenario. In other words, given a real-life situation of connecting a number of locally installed networking devices, how does one go about designing a solution to this problem? What are the choices to be made? What are the key issues and trends? What are the alternatives?

Chapter 8 focuses on the basic building blocks of LANs, namely devices such as repeaters, bridges, routers, brouters, gateways, hubs, file servers, and LAN switches. For each device, we will briefly review their operation, focusing upon how they process frames, because such processing has an integral relationship to network design and design constraints.

In Chapter 9, we will focus our attention to the topic of Ethernet LAN topology. Thus, our focus will include star- and bus-based Ethernet, as well as the use of single and tiered switches. In examining switches, we will also look at the placement of so-called server farms and note why literally placing all of your eggs in one basket may not be suitable for many organizations.

It is estimated that the great majority (and increasing number) of installed LANs are Ethernet-based vis-à-vis Token Ring and asynchronous transfer mode (ATM). One of the reasons for this may be the significantly less expensive hardware, with acceptable performance at the same time. Other arguments may also be made for the fact that Token Ring/ATM LANs have become rather like the dodo bird, and so in this edition we specifically exclude their discussion. Also in this chapter, we will briefly discuss two evolution mechanisms for LAN selection within an organization — bottom up and top down. In the former, a three-tiered architecture results, whereas in the latter, an enterprise LAN strategy drives a centralized approach to the problem.

Chapter 10 is a tutorial on the Ethernet family. Commencing with a review of the various versions of the 10BASE family of networks, we move on to the Fast Ethernet family, the Gigabit family, and the recently standardized 10 Gbps version of Ethernet. We will look at the constraints associated with each member of the Ethernet family, including the so-called 5-4-3 rule and cabling limitations. We will also examine transmission media and their characteristics, including twisted pair, coax, and both single mode and multimode fiber optics.

In Chapter 11, we turn our attention to Ethernet performance characteristics. Topics in this chapter include the various issues at the data link layer, including framing, the interframe gap, frame overhead, and their effect on performance and information transfer, as well as reliable data exchange and error management.

Chapter 12 is concerned with issues at the network, transport, and application layers with a further look at frames and their effect on processing, delays, and latency considerations. Because the Transmission Control Protocol/Internet Protocol (TCP/IP) is by far the most popular protocol transported by Ethernet, we will note the delays required when packets are transported within an Ethernet frame. In doing so, we can determine if it is practical to transport delay sensitive information, such as voice over an Ethernet network.

No discussion of LANs would be complete without a discussion of wireless LANs. In Chapter 13, we provide an overview of the Institute for Electrical and Electronic Engineering (IEEE) standards — IEEE 802.11a, IEEE 802.11b, and IEEE 802.11g. In addition, we will examine the limitations of access points with respect to client support and security, noting how recently developed wireless switches can enhance both management and security.

No LAN is an island. There is always the ineluctable requirement of hooking up your LAN to other networks, either locally to another LAN, to the Internet, or by a WAN to another LAN. Chapter 14 deals with internetworking and the problems associated with interconnecting geographically separated LANs. In concluding this chapter, we note that in certain quarters, network management is regarded as partly a design issue. Our position is that this is a modification tool, as the network needs to be up and running in the first place to deploy network management techniques. Accordingly, we will exclude a detailed discussion of this topic in this book.

I

WIDE AREA NETWORK
DESIGN TECHNIQUES

This section consists of six chapters that primarily focus upon the design of WANs. As indicated in the first chapter in this book, because some network design techniques apply to both WANs and LANs, we decided where to place certain chapters based upon the primary use of the techniques described in those chapters. However, when material applies to both types of networks, a description of the use of the techniques for WAN and LAN design is included in the chapter.

This section includes six chapters, commencing with Chapter 2, which focuses on transmission characteristics. In that chapter, we will examine communications constraints in the form of throughput, response time, and various delay factors. We will also construct several models to illustrate the effect of errors upon transmission. Although the primary orientation of the chapters in this section is toward WAN design techniques, it should be obvious from our short description of the contents of Chapter 2 that throughput and response time are also constraints associated with LANs. Thus, at applicable points in each chapter in this section, we will note how we can use certain material presented for the design of WANs for their LAN cousins.

2

TRANSMISSION CHARACTERISTICS

In this chapter, we will examine the characteristics of batch and interactive transmission as all communications applications can be placed into one of these two categories. To accomplish this, we will first review the constraints associated with the development of communications applications and the relationship of those constraints to each transmission category. Next, we will develop several models that will be used to project performance. Although we will limit our modeling to one protocol, its creation and exercise will provide a foundation for developing similar models for other transmission protocols.

2.1 COMMUNICATIONS CONSTRAINTS

The development of communications-based applications that are the foundation of our modern society involves many trade-offs in terms of the use of different types of communications facilities, types of terminal devices, hours of operation, and other constraints. Two of the key constraints associated with the development of communications applications are throughput and response time.

2.1.1 Throughput

Throughput is a measurement of the transmission of a quantity of data per unit of time, such as the number of characters, records, blocks, or print lines transmitted during a predefined interval. Throughput is normally associated with batch systems where the transmission of a large volume of data to a distant location occurs for processing, file updating, or printing.

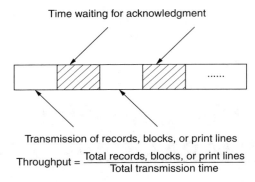

Figure 2.1   Batch Transmission and Throughput

As this is typically an extension of batch processing and because it occurs remotely from a data center, the device the transmission is from or to is referred as a remote batch or remote job entry device.

In most batch transmission systems, a group of data representing a record, block, or print line is transmitted as an entity. Its receipt at its destination must be acknowledged when using many protocols prior to the next grouping of data being transmitted. Figure 2.1 illustrates the operation of a batch transmission system by time, with the waiting time indicated by shaded areas. Because throughput depends upon the waiting time for acknowledgments of previously transmitted data, one method used to increase throughput is to transmit more data prior to requiring an acknowledgment.

2.1.2   Response Time

Response time is associated with communications where two entities interact with one another, such as a terminal user entering queries into a computer system. Here, each individual transaction or query elicits a response and the time taken to receive the response is of prime importance.

Response time can be defined as the time between a query being transmitted and the receipt of the first character of the response to the query. Figure 2.2 illustrates interactive transmission response time.

The optimum response time for an application depends upon the type of application. For example, a program that updates an inventory could have a slower response time than an employee badge reader or an airline reservation system. The reason for this is that an employee entering information from a bill of lading or other data that is used to update a firm's inventory would probably find a five- or ten-second response time

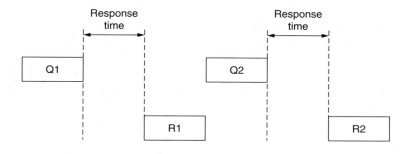

Figure 2.2   Interactive Transmission Response Time

to be satisfactory. For a badge reader system where a large number of workers arrive and leave during a short period of time, queues would probably develop if the response time was similar. With regard to airline reservation systems, many potential customers require a large amount of information concerning discount prices, alternate flights, and time schedules. If the airline reservation clerk experienced a slow response time in scrolling through many screens of information to answer a customer query, the cumulative effect of a five-second response time could result in the customer hanging up in disgust and calling a competitor. For other interactive communication applications, such as automated teller machines, competitive advertising has almost made slow response an issue involving the violation of a user's fundamental rights. In certain locations, banks battle against one another in advertisements over which one has the fastest teller machines. This is yet another example of the use of communications to gain a competitive position.

Although we typically think of throughput and response time as WAN design metrics, they also have a role in LAN design. That is, the location of servers with respect to other servers and client LAN stations can cause bottlenecks that affect the transfer of information. In addition, the type of server processor, its network interface card, and the level of utilization of the server and LAN all have a bearing on throughput and response time. Because many times it is far easier to measure throughput and response time than a series of metrics, some persons elect to alter a network configuration and record easy to measure metrics rather than individual metrics.

2.2   INFORMATION TRANSFER RATE

Although it is common for many persons to express interest in the data transmission rate of a communications application, relying upon this rate can be misleading. In actuality, it is more important that analysts and designers focus their attention upon the transfer rate of information in bits (TRIB). This rate, which is normally measured in bits per second (bps), is similar to a data transmission rate and concerns the rate of flow of actual informative data. In comparison, a data transmission rate is the rate at which bits flow across a transmission medium where the bits can represent both informative data and control information. In this section, we will examine several factors that affect the TRIB that will provide the foundation for developing several transmission models later in this chapter.

2.2.1 Delay Time

When data is transmitted between terminals, a terminal and a computer, or two computers, several delay factors may be encountered that cumulatively affect the information transfer rate. Data transmitted over a transmission medium must be converted into an acceptable format for that medium. When digital data is transmitted over analog telephone lines, modems must be employed to convert the digital pulses of the business machine into a modulated signal acceptable for transmission on the analog telephone circuit. Even when transmission occurs on an all-digital facility, other equipment delay times will affect the information transfer rate. Here, data service units (DSUs) used for transmission at data rates up to 56/64 kilobits per second (kbps) or channel service units (CSUs) used for T-carrier transmission at 1.544 megabits per second (Mbps) will require some time to convert unipolar digital signals into bipolar digital signals.

The time between the first bit entering a modem or service unit and the first modulated signal produced by the device is known as the modem's or service unit's internal delay time. Because two such devices are required for a point-to-point circuit, the total internal delay time encountered during a transmission sequence equals twice the device's internal delay time. Such times can range from a few to ten or more milliseconds (ms).

A second delay encountered on a circuit is a function of the distance between points to be connected and is known as the circuit or propagation delay time. This is the time required for the signal to be propagated or transferred down the line to the distant end. Propagation delay time can be approximated by equating 1 ms to every 150 circuit miles and adding 12 ms to the total.

2.2.2 Other Delay Factors

Once data is received at the distant end, it must be acted upon, resulting in a processing delay that is a function of the computer or terminal employed, as well as the quantity of transmitted data that must be acted upon. Processing delay time can range from a few milliseconds, where a simple error check is performed to determine if the transmitted data was received correctly, to many seconds, where a search of a database must occur in response to a transmitted query. Each time the direction of transmission changes in a typical half-duplex protocol, control signals at the associated modem-to-computer and modem-to-terminal interfaces change. The time required to switch control signals to change the direction of transmission is known as line turnaround time and can result in delays of up to 250 or more milliseconds, depending upon the transmission protocol employed.

We can denote the effect of the previously mentioned delays upon the information transfer rate by modeling the binary synchronous (BISYNC) communications protocol and a few of its derivatives. Although the resulting models are primarily applicable to batch transmission, the differences between the half- and full-duplex models will illustrate the advantages of the latter; similar gains can be expected from full-duplex transmission based upon the use of protocols that support this method of data transmission.

2.3   BISYNC COMMUNICATIONS PROTOCOL MODELS

At one time, one of the most commonly employed transmission protocols was the BISYNC communications control structure. This line control structure was introduced in 1966 by International Business Machine Corporation (IBM) and was used for transmission by many medium- and high-speed devices to include terminal and computer systems. BISYNC provides a set of rules that govern the synchronous transmission of binary-coded data. Although this protocol could be used with a variety of transmission codes, it is normally limited to a half-duplex transmission mode and requires acknowledgment of the receipt of every block of transmitted data. In an evolutionary process, a number of synchronous protocols were developed to supplement or serve as a replacement to BISYNC, the most prominent being the High-Level Data-Link Control (HDLC) protocol defined by the International Standards Organization (ISO). Although BISYNC represents an obsolete protocol, its modeling provides an easy method to illustrate the efficiency of data transfer as the quantity of data in a block varies. Thus, the original selection of the BISYNC protocol in the first edition of this book remains a valid selection for future editions.

The key difference between BISYNC and HDLC protocols is that BISYNC is a half-duplex, character-oriented transmission control structure and HDLC is a bit-oriented, full-duplex transmission control structure. We can investigate the efficiency of these basic transmission control structures and the effect of different delays upon their information transfer efficiency. To do so, an examination of some typical error control procedures is first required.

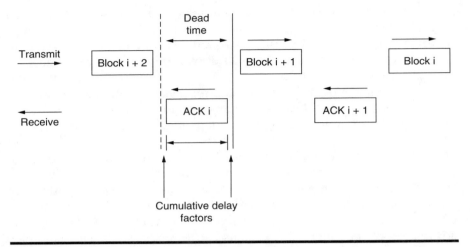

Figure 2.3   Stop and Wait ARQ

2.3.1   Error Control

The most commonly employed error control procedure is known as automatic repeat request (ARQ). In this type of control procedure, when an error is detected, the receiving station requests that the sending station retransmit the message. Two types of ARQ procedures have been developed:

1. Stop and wait ARQ
2. Go back in ARQ (or continuous ARQ)

Stop and wait ARQ is a simple type of error control procedure. Here, the transmitting station stops at the end of each block and waits for a reply from the receiving terminal pertaining to the block's accuracy (ACK) or error (NAK) prior to transmitting the next block. In this type of error control procedure (see Figure 2.3), the receiver transmits an acknowledgment after each block. This can result in a significant amount of cumulative delay between data blocks. Here, the time between transmitted blocks is referred to as dead time, which acts to reduce the effective data rate on the circuit. When the transmission mode is half-duplex, the circuit must be turned around twice for each block transmitted, once to receive the reply (ACK or NAK) and once again to resume transmitting. These line turnarounds, as well as such factors as the propagation delay time, station message processing time, and the modem or service set internal delay time, all contribute to what is shown as the cumulative delay factors.

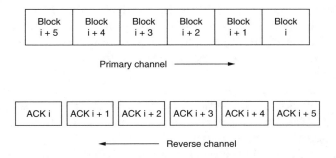

Figure 2.4   Go Back in ARQ

When the go back in ARQ type of error control procedure is employed, the dead time can be substantially reduced to the point where it may be insignificant. One way to implement this type of error control procedure is to use a simultaneous reverse channel for acknowledgment signaling as illustrated in Figure 2.4. In this type of operating mode, the receiving station sends back an ACK or NAK response on the reverse channel for each transmitted block. If the primary channel operates at a much higher data rate than the reverse channel, many blocks may be received prior to the transmitting station receiving the NAK in response to a block that the receiving station found in error. The number of blocks one may go back to request a retransmission, $n,$ is a function of the block size and buffer area available in the business machines and terminals at the transmitting and receiving stations, the ratio of the data transfer rates of the primary and reverse channels, and the processing time required to compute the block check character and transmit an acknowledgment. For the latter, this time is shown as small gaps between the ACK and NAK blocks in Figure 2.4.

In a go back in ARQ error control procedure, the transmitter continuously sends messages until the receiver detects an error. The receiver then transmits a negative acknowledgment on the reverse channel and the transmitter retransmits the block received in error. Some versions of this technique require blocks sent before the error indication was encountered to be retransmitted in addition to the block received in error.

2.3.2   Half-Duplex Throughput Model

When a message block is transmitted in the BISYNC control structure, a number of control characters are contained in that block in addition to the message text. If the variable C is assigned to represent the number of control characters per block and the variable D is used to represent the number of data characters, then the total block length is C + D. If the data transfer rate expressed in bps is denoted as T_R and the number of bits per character is denoted as B_c, then the transmission time for one character is equal to B_c/T_R, which can be denoted as T_C. Because D + C characters are contained in a message block, the time required to transmit the block will become $T_C \times (D + C)$. Once the block is received, it must be acknowledged. To do so, the receiving station is required to first compute a block check character (BCC) and compare it with the transmitted BCC character appended to the end of the transmitted block. Although the BCC character is computed as the data is received, a comparison is performed after the entire block has been received and only then can an acknowledgment be transmitted. The time to check the transmitted and computed BCC characters and form and transmit the acknowledgment is known as the processing and acknowledgment time (T_{PA}).

When transmission is half-duplex, the line turnaround time (T_L) required to reverse the transmission direction of the line must be added. Normally, this time includes the Request-to-Send/Clear-to-Send (RTS/CTS) modem delay time as well as each of the modems' internal delay times. For the acknowledgment to reach its destination it must propagate down the circuit and this propagation delay time, denoted as T_p, must also be considered. If the acknowledgment message contains A characters then, when transmitted on the primary channel, $A \times B_c/T_R$ seconds are required to send the acknowledgment.

Once the original transmitting station receives the acknowledgment, it must determine if it is required to retransmit the previously sent message block. This time is similar to the processing and acknowledgment time previously discussed. To transmit either a new message block or repeat the previously sent message block, the line must be turned around again and the message block will require time to propagate down the line to the receiving station. Thus, the total time to transmit a message block and receive an acknowledgment, denoted as T_B, becomes:

$$T_B = T_C \times (D + C) + 2 \times (T_{PA} + T_L + T_p) + (A \times B_C/T_R)$$

Because efficiency is the data transfer rate divided by the theoretical data transfer rate, the transmission control structure efficiency (E_{TCS}) becomes:

$$E_{TCS} = \frac{B_C \times D \times (1 - P)}{T_R \times T_B}$$

where P is the probability that one or more bits in the block are in error, causing a retransmission to occur.

Although the preceding is a measurement of the transmission control structure efficiency, it does not consider the data code efficiency, which is the ratio of information bits to total bits per character. When the data code efficiency is included, we obtain a more accurate measurement of the information transfer efficiency. We can call this ratio the information transfer ratio (ITR), which will provide us with a measurement of the protocol's information transfer efficiency. This results in:

$$ITR = \frac{B_{IC} \times E_{TCS}}{B_C} = \frac{B_{IC} \times D \times (1 - P)}{T_R \times T_B}$$

Where,
ITR = information transfer ratio
B_{IC} = information bits per character
B_C = total bits per character
D = data characters per message block
A = characters in the acknowledgment message
C = Control characters per message block
T_R = data transfer rate (bps)
T_C = transmission time per character (B_C/T_R)
T_{PA} = processing and acknowledgment time
T_L = line turnaround time
T_P = propagation delay time
P = probability of one or more errors in block

From the preceding, the ITR provides us with a more accurate measurement of the efficiency of the transmission control structure.

2.3.2.1 Computation Examples

We will assume that our data transmission rate is 4800 bps and that we will transmit information using a BISYNC transmission control structure employing a stop and wait ARQ error control procedure. Furthermore, let us assume the following parameters:

A = 4 characters per acknowledgment
B_{IC} = 8 bits per character
B_C = 8 bits per character
D = 80 data characters per block
C = 10 control characters per block
T_R = 4800 bps

T_C = 8/4800 = 0.00166 s per character
T_{PA} = 20 ms = 0.02 s
T_L = 100 ms = 0.10 s
T_P = 30 ms = 0.03 s
P = 0.01

Then,

$$ITR = \frac{8 \times 80 \times (1 - 0.01)}{4800 \times \left[0.00166 \times (80 + 10) + 2 \times (0.02 + 0.03 + 0.1) + (4 \times 8/4800) \right]} = 0.28905$$

Because the TRIB is equal to the product of the data transfer rate and the efficiency of the line discipline or protocol that we have denoted as the information transfer ratio, we obtain:

$$TRIB = ITR \times T_R = 0.28905 \times 4800 = 1387 \, bps$$

For the preceding example, approximately 29 percent of the data transfer rate (1387/4800) is effectively used.

Let us now examine the effect of doubling the text size to 160 characters while the remaining parameters except P continue as before. Because the block size has doubled, P approximately doubles, resulting in the ITR becoming:

$$ITR = \frac{8 \times 160 \times (1 - 0.02)}{4800 \times \left[0.00166 \times (160 + 10) + 2 \times (0.02 + 0.03 + 0.1) + (4 \times 8/4800) \right]} = 0.44294$$

With an ITR of 0.44294 the TRIB now becomes:

$$TRIB = ITR \times T_R = 0.44294 \times 4800 = 2156 \, bps$$

Here, doubling the block size raises the percentage of the data transfer rate effectively used to 44.92 percent from approximately 29 percent.

To assist in the tabulation of ITRs for increasing block sizes, a computer program was written using Microsoft® Quick Basic. Table 2.1 contains a

Table 2.1 **BISYNC.BAS** Program Listing

```
REM Half Duplex Bysinc transmission model
DIM ITR(40), D(40)
A = 4'characters per acknowledgement
BIC = 8 'information bits per character
BC = 8   'total bits per character
D = 80' data characters per block
C = 10'control characters per block
TR = 4800'transmission rate in bits per second (bps)
TC = BC / TR'transmission time per character
TPA = .02 'processing and acknowledgement time in seconds (200
milliseconds)
TL = .1 'line turnaround time in seconds (100 milliseconds)
TP = .03' propagation dely time in seconds (30 milliseconds)
P = .01'probability one or more bits in block in error
REM vary the block size from 40 to 2400 characters by 40
CLS
K = 1'initialize array pointer
PRINT "ITR        BLOCK SIZE            ITR          BLOCK SIZE"
FOR D = 80 TO 3200 STEP 80
ITR = (BIC * D * (1 - P)) / (TR * (TC * (D + C) + 2 *
(TPA + TL + TP) + (A * BC / TR)))
ITR(K) = ITR
D(K) = D
K = K + 1       'increment array pointer
P = P + .01    'increment probability of bit error
NEXT D
FOR K = 1 TO 20
PRINT USING "#.#####    ####     #.#####   ####"; ITR(K);
PRINT D(K), ITR(K + 20), D(K + 20)
NEXT K
END
```

program listing of the file named BISYNC.BAS. This program uses the previously assumed transmission control structure parameters to compute the information transfer ratio as the block size was varied from 80 to 3200 characters in increments of 80 characters. Table 2.2 contains the results obtained from the execution of the BISYNC.BAS program. In examining this table, you will note that the ITR increases as the block size increases until a block size of 1040 characters is reached. Thereafter, the ITR decreases as the block size increases. This indicates that as the block size increases with a constant error rate, a certain point is reached where the time required to retransmit a long block every so often, due to one or more bits in the block being received in error, negates the efficiencies obtained by the enlargement of the block size. For the parameters used in the previously constructed model, the optimum block size is 1024 characters.

Table 2.2 Information Transfer Ratio versus Block Size (Probability of Block Error = 0.01)

ITR	BLOCK SIZE	ITR	BLOCK SIZE
0.28905	80	.7082177	1680
0.44294	160	.7025588	1760
0.53641	240	.6965585	1840
0.59767	320	.6902554	1920
0.63973	400	.6836827	2000
0.66944	480	.6768689	2080
0.69072	560	.6698385	2160
0.70600	640	.662613	2240
0.71685	720	.6552108	2320
0.72435	800	.6476484	2400
0.72924	880	.6399402	2480
0.73206	960	.6320988	2560
0.73323	1040	.6241356	2640
0.73303	1120	.6160604	2720
0.73171	1200	.6078825	2800
0.72944	1280	.5996097	2880
0.72638	1360	.5912493	2960
0.72264	1440	.5828077	3040
0.71832	1520	.574291	3120
0.71349	1600	.5657043	3200

In actuality, a bit error rate of 1×10^{-5} or better is usually experienced on dial-up switched network facilities. Because the initial block size in the BISYNC.BAS program is 80 characters, this equates to a transmission of 640 data bits. Thus, a more realistic measurement of the ITR could be obtained by setting the probability of a bit error in a block (P) to 0.0064 in the program and incrementing that probability by 0.0064 for each block increase of 80 characters or 640 data bits. Table 2.3 contains the results obtained by modifying the BISYNC.BAS program as previously discussed. Note that the optimum block size has increased to 1360 characters.

2.3.2.2 Block Size versus Error Rate

A comparison of the data contained in Table 2.1 and Table 2.2 illustrates an important concept that warrants elaboration as this concept has been applied to a variety of transmission systems, including the Microcom Networking Protocol (MNP) used in many modems. The concept is that an optimum block size is inversely proportional to the error rate. That is, a lower error rate results in a lower optimum block size.

Under one of the classes of the MNP protocol, interactive data is assembled into packets for transmission. The size of these packets is

**Table 2.3 Information Transfer Ratio versus Block Size
(Probability of Block Error = 0.064)**

ITR	BLOCK SIZE	ITR	BLOCK SIZE
0.28905	80	.7727641	1680
0.44456	160	.770653	1760
0.54039	240	.7682045	1840
0.60439	320	.7654569	1920
0.64943	400	.762443	2000
0.68226	480	.7591909	2080
0.70676	560	.7557247	2160
0.72533	640	.7520657	2240
0.73954	720	.7482322	2320
0.75042	800	.7442405	2400
0.75873	880	.7401047	2480
0.76501	960	.7358373	2560
0.76963	1040	.7314495	2640
0.77292	1120	.7269512	2720
0.77509	1200	.7223513	2800
0.77634	1280	.7176577	2880
0.77679	1360	.7128776	2960
0.77658	1440	.7080172	3040
0.77579	1520	.7030826	3120
0.77449	1600	.6980789	3200

adaptive and is based indirectly upon the line error rate. We say the adaptive packets are based indirectly upon the line error rate because the protocol counts negative acknowledgments in place of performing a bit error rate test that would preclude the transmission of data. In this instance, Microcom took advantage of the fact that the error rate can vary considerably over the switched telephone network. Thus, when a lower error rate occurs the protocol places more data into a block or packet, but a higher error rate results in the protocol reducing the size of the packet.

To obtain another view of the relationship between protocol efficiency, block size, and error rate the previously constructed Basic program was executed using a value of zero (0.0) for P. This is the ideal situation where a continuously increasing block size produces additional efficiencies. Table 2.4 lists the ITR as a function of the block size when the probability of an error occurring is set to zero. Although the ITR will approach a value of unity, it will do so at an extremely large block size, which would be impractical to effect due to finite buffer sizes in computers.

In Figure 2.5, the ITR was plotted as a function of block size for the 0.01 and 0.0064 probability of error conditions as well as the error-free condition. The top line represents a zero probability of error, which approaches unity as the block size increases. Because an error-free line

**Table 2.4   Information Transfer Ratio versus Block Size
(Probability of Block Error = 0.0)**

ITR	BLOCK SIZE	ITR	BLOCK SIZE
0.28905	80	.8875133	1680
0.44746	160	.8917093	1760
0.54747	240	.8955752	1840
0.61634	320	.8991485	1920
0.66667	400	.9024612	2000
0.70504	480	.9055409	2080
0.73528	560	.9084112	2160
0.75971	640	.9110928	2240
0.77987	720	.9136038	2320
0.79678	800	.9159599	2400
0.81117	880	.918175	2480
0.82357	960	.9202614	2560
0.83436	1040	.9222301	2640
0.84384	1120	.9240906	2720
0.85222	1200	.9258517	2800
0.85970	1280	.9275211	2880
0.86641	1360	.9291059	2960
0.87246	1440	.9306122	3040
0.87795	1520	.9320458	3120
0.88294	1600	.9334119	3200

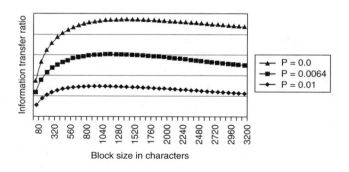

Figure 2.5   ITR Plotted against Block Size

is not something a data transmission network designer or analyst can expect, there will always exist a maximum block size beyond which line efficiency will decrease as indicated by the middle and lower curves in Figure 2.5. We can take advantage of this known phenomenon by either adjusting the block size of batch transmission to correspond to average line conditions or we can employ an adaptive protocol similar to the previously described MNP protocol. Regardless of the method used, it is important to understand that, in many instances, software default settings supplied with a vendor's product may be designed for average line conditions. As such, they may be far from optimum for the facilities used by your organization and a little effort or experimentation in changing the default settings may significantly increase the transmission efficiency of the product.

Now that we have examined how the size of a data block or packet can be altered to improve communication efficiency, let us focus our attention upon the effect of varying the protocol. Let us first examine the use of a return or secondary channel to transmit acknowledgments and then investigate the efficiency obtained from the use of a full-duplex protocol.

2.3.3   Return Channel Model

A return or secondary channel is built into many synchronous modems and provides a mechanism for the transmission of acknowledgments in one direction while data flows in the opposite direction. The use of this return channel eliminates the necessity of line turnarounds; however, transmission is still half-duplex when a stop and wait ARQ error control procedure is used, because an acknowledgment is only transmitted after each received message block is processed. However, because acknowledgments can simultaneously flow in the reverse direction, this model simulates a partial full-duplex transmission method.

When the message block is sent to the receiving station, both propagation delay and processing delay are encountered. When the acknowledgment is returned, one additional propagation delay and processing delay results. In addition to these delays, one must also consider the time required to transmit the acknowledgment message. If A denotes the length in characters of the acknowledgment message and T_S is the reverse channel data rate in bps, then the transmission time for the acknowledgment becomes $(A \times B_C)/T_S$. The total delay time due to the propagation and processing, as well as the acknowledgment transmission time, becomes:

$$2 \times \left(T_{PA} + T_P\right) + \frac{A \times B_C}{T_S}$$

Thus, the ITR becomes:

$$ITR = \frac{B_{IC} \times D_1 \times \left(1 - P\right)}{T_R \times \left[T_C \times \left(D + C\right) + 2 \times \left(T_{PA} + T_P\right) + A \times B_C / T_S\right]}$$

Let us examine the effect of this modified transmission procedure on the previous example where data was packed 80 characters per block. Let us assume that a 75-bps reverse channel is available and our acknowledgment message comprises four 8-bit characters. Then:

$$\text{ITR} = \frac{8 \times 80 \times (1 - 0.01)}{4800 \times \left[0.00166 \times (80 + 10) + 2 \atop \times (0.02 + 0.03) + (4 \times 8/75) \right]} = 0.1953$$

Note that the ITR actually decreased. This was caused by the slowness of the reverse channel where it took $0.4266 \times (4 \times 8/75)$ seconds to transmit an acknowledgment. In comparison, the two line turnarounds that were eliminated only required 0.2 seconds when the acknowledgment was sent at 4800 bps on the primary channel. This modified procedure is basically effective when the line turnaround time exceeds the transmission time of the acknowledgment on the return channel. Thus, for a reverse channel to be efficient you should ensure that:

$$\frac{A \times B_C}{T_S} < 2 \times T_L$$

Although the original edition of this book predated the advent of DSL technology where the bandwidth is partitioned asymmetrically, the previous model illustrates why downloads from the Internet occur faster than if you use your DSL connection to upload a file. That is, the asymmetrical bandwidth provides a higher operating download channel than upload channel.

2.3.4 Full-Duplex Model

A much greater throughput efficiency with the stop and wait ARQ error control procedure can be obtained by employing a full-duplex mode of transmission. Although this requires a four-wire circuit or communications devices that split a communications channel into two separate paths by frequency, the communications devices and line do not have to be reversed. This permits an acknowledgment to be transmitted at the same data rate as the message block, but in the reverse direction without the line turnaround. Thus, the ITR becomes:

$$\text{ITR} = \frac{B_{IC} \times D \times (1 - P)}{T_R \times \left[T_C \times (D + C) + 2 \times (T_{PA} + T_P) \right]}$$

Again, returning to the original 80-character block example, we obtain:

$$ITR = \frac{8 \times 80 \times \left(1 - 0.01\right)}{4800 \times \left[0.00166 \times \left(80 + 10\right) + 2 \times \left(0.02 + 0.03\right)\right]} = 0.528$$

Note that the ability to simultaneously transmit data blocks in one direction and acknowledgments in the opposite direction raised the ITR from 0.28905 to 0.528. Although the preceding model execution indicates that a full-duplex protocol can be much more efficient than a half-duplex protocol, it is also important to understand that the overall efficiency is highly related to the size of the data blocks being transmitted. As an example, an interactive protocol in which the data block size averaged 20 characters would have an ITR of 0.22 when transmission was full-duplex. Although this would be considerably higher than a half-duplex protocol in which the ITR would be 0.0925 with a block size of 20 characters, it is still relatively inefficient. This also explains why routers, multiplexers, and control units that gather and process data from several sources into a larger data block prior to transmission increase the efficiency of transmission.

To conclude this chapter, let us examine a variation of the previous model to examine the effect of a go back in ARQ error control procedure. This error control procedure is employed in some nonstandard BISYNC protocols and is a popular option supported by HDLC. Under a go back in ARQ error control procedure only the block received in error is retransmitted. Here, the ITR becomes:

$$ITR = \frac{B_{IC} \times \left(1 - P\right)}{T_R \times \left[T_C \times \left(D + C\right)\right]}$$

Again, substituting values from the original example we obtain:

$$ITR = \frac{8 \times 80 \times \left(1 - 0.01\right)}{4800 \times \left[0.00166 \times \left(80 + 10\right)\right]} = 0.8835$$

This is obviously the most efficient technique because the line turnaround is eliminated and the processing and acknowledgment time (T_{PA}) and propagation delay time (T_P) in each direction are nullified due to simultaneous message block transmission and acknowledgment response. Based upon the preceding, it becomes quite clear why a full-duplex transmission protocol that supports selective rejection is the most efficient method of data transmission.

3

WIDE AREA NETWORK LINE FACILITIES AND ECONOMIC TRADE-OFFS

One of the most common problems facing network analysts and designers involves the selection of an appropriate line facility to satisfy the data transmission requirements of their organization. In this chapter, we will examine the three basic types of line connections you can consider to support the WAN transmission requirements of your organization. Using this information as a base, we will then examine several types of analog and digital transmission facilities as well as hybrid analog line facilities, including foreign exchange (FX) and wide area telecommunications service (WATS). Because tariffs, like taxes, constantly change, we will employ a set of rate schedules only to illustrate the economic trade-offs associated with different comparisons between two or more line facilities. Although the rate schedules presented in this chapter may be similar to future tariffs, you should contact communications carriers to determine the actual tariffs in effect when they perform their economic analysis. Regardless of the actual tariffs in effect when your analysis is performed, the methods used in performing the economic comparisons presented in this chapter can be used as a guide for your analysis.

3.1 LINE CONNECTIONS

There are three basic types of line connections available to connect terminal devices to computers or to other terminals over distances that normally preclude the use of a LAN:

1. Dedicated lines
2. Switched lines
3. Leased lines

3.1.1 Dedicated Line

A dedicated line is similar to a leased line in that the terminal is always connected to the device on the distant end, transmission always occurs on the same path, and, if required, the line can be easily tuned to increase transmission performance.

The key difference between a dedicated and a leased line is that a dedicated line refers to a transmission medium internal to a user's facility, where the customer has the right of way for cable laying, whereas a leased line provides an interconnection between separate facilities. The term facility is usually employed to denote a building, office, or industrial plant. Dedicated lines are also referred to as direct connect lines and normally link a terminal or business machine on a direct path through the facility to another terminal or computer located at that facility. The dedicated line can be a wire conductor installed by the employees of a company or by the computer manufacturer's personnel. In certain situations, a dedicated line can even represent a line facility installed by a communications carrier to interconnect two buildings on a corporate campus, where an organization pays a onetime fee and has exclusive use of the transmission capability of the line.

Normally, the only cost associated with a dedicated line in addition to its installation cost is the cost of the cable required to connect the devices that are to communicate with one another. Thus, the cost associated with a dedicated line can be considered as a onetime charge. Once installed, the dedicated line can be expected to have no associated recurring cost except maintenance, which may occur randomly in time. In comparison, facilities obtained from communications carriers that use the carrier's plant normally have both onetime and recurring costs. The onetime cost is for the installation of the facility. The recurring cost can include a small monthly access line fee as well as a varying charge based upon minutes of use for communications over the switched telephone network. For an analog leased line, the recurring cost can include a monthly fee for line conditioning as well as the monthly cost of the portions of the line furnished by local exchange and interexchange carriers. The relationship between local exchange and interexchange carriers will be described when we examine leased lines in this section.

3.1.2 Leased Line

A leased line is commonly called a private line and is obtained from a communications company to provide a transmission medium between two facilities, which could be in separate buildings in one city or in distant cities. In addition to a onetime installation charge, the communications carrier will normally bill the user on a monthly basis for the leased line, with the cost of the line usually based upon the distance between the locations to be connected.

In the United States, until 1984, prior to AT&T's divestiture, the computation of the cost of a leased line was fairly simple as there were only two types of tariffs to consider in the United States — interstate and intrastate. AT&T and other common carriers, such as MCI and Sprint, filed interstate tariffs with the Federal Communications Commission (FCC). Intrastate tariffs were filed by Bell and independent operating companies with the public utility commissions of the states they operated in.

Although the distinction between interstate and intrastate survived in the United States after divestiture, a new tariff criteria was added. This criteria considers whether or not a service is located within the local areas served by the divested local operating companies. These areas are called local access and transport areas (LATAs) and essentially correspond to the metropolitan statistical areas defined by the U.S. Commerce Department.

When a service is entirely within a LATA, it is an intra-LATA service, although a service linking two or more LATAs is known as an inter-LATA service. Because LATAs can cross state boundaries or reside entirely within a state, there are both interstate and intrastate tariffs for inter- and intra-LATA service.

Within each LATA are interface points called points of presence (POPs) that are by law the only locations where interexchange carriers (IEC) can receive and deliver traffic. Thus, to establish a service between LATAs, the local exchange carrier (LEC) facilities can be used to provide a connection to a selected IEC's POP. Once an IEC is selected, they will route their service to a POP at the destination LATA where the same or a different LEC provides a connection from the POP to the customer's distant premises.

Because many LATAs cross state boundaries while other LATAs are located entirely within a state, users have to contend with up to six types of tariffs in place of two prior to divestiture. Today there are inter-LATA, intra-LATA, and LATA access tariffs for both interstate and intrastate. To add to the options available for user consideration, by law, users have the option to bypass the LEC and construct their own connections to the nearest central office of an interexchange carrier.

To illustrate the cost components of a leased line consider Figure 3.1, which illustrates the relationship between LECs and an IEC. The POP or interface points where the IEC connects to the LEC is normally a building housing facilities of both the IEC and LEC.

The distinction between the local exchange carrier and interexchange carrier was prominently in the news during the summer of 2003. In a series of articles it was reported that MCI, formerly known as WorldCom, which committed an $11 billion fraud, was routing telephone calls such

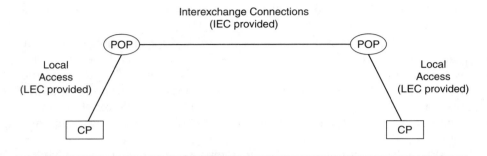

Figure 3.1   Leased Line Components
Unless the LEC is bypassed, a leased line consists of two relatively short access lines provided by the same or different LECs and a line from the interexchange carrier used to connect each access line at a POP.

that the origination number that indicated a long-distance call was changed to a local number or blanked out, allowing MCI to reduce its payment to the LEC for completing the call. Apparently, WorldCom was able to shave a nickel here and a nickel there until it avoided payments of hundreds of millions of dollars to LECs.

Returning our attention to Figure 3.1, users can select one of two methods to establish a leased line. They can order:

1. Transport circuit, also known as a baseline service
2. Total-service circuit

When a transport service is ordered, the leased line is obtained from the IEC as a connection between the two distant central offices of that carrier, with the user responsible for coordinating with the LEC for the local connections at each end. Under this arrangement, the user is responsible for ordering, maintaining, and testing the entire end-to-end connection, as well as for overseeing the actual installation. When this occurs, the user will receive bills from each LEC for local access to the POP, as well as from the IEC for the long-distance circuit.

Under the total-service concept, the IEC has end-to-end responsibility for ordering, installing, servicing, and billing, as well as for any maintenance. The IEC will coordinate the installation of the local access lines to the POP with each LEC. Normally, the IEC adds a surcharge for the extra service and bills the user directly for both access lines and the IEC circuit on one bill.

The availability of a particular type of service depends upon both the LEC and IEC. In some instances, the LEC may not provide a service that the IEC offers. One example would be fractional T1 (FT1), where an LEC might require the user to obtain a T1 access line to an FT1 facility. Similarly, not every IEC service may be available at each POP.

3.1.3 Switched Line

A switched line, often referred to as a dial-up line, permits contact with all parties having access to the public switched telephone network (PSTN). If the operator of a terminal device wants access to a computer, he or she dials the telephone number of a telephone that is connected to the computer. In using switched or dial-up transmission, telephone company switching centers establish a connection between the dialing party and the dialed party. After the connection has been set up, the terminal and the computer conduct their communications. When communications have been completed, the switching centers disconnect the path that was established for the connection and restore all paths used so they become available for other connections.

The cost of a call on the PSTN varies by country. In the United States, the cost of a call was for many years based upon many factors, including the time of day when the call was made, the distance between called and calling parties, the duration of the call, and whether or not operator assistance was required in placing the call. Direct dial calls made from a residence or business telephone without operator assistance are billed at a lower rate than calls requiring operator assistance. In addition, most telephone companies originally had three categories of rates:

1. Day
2. Evening
3. Night and weekend

Calls made between 8 a.m. and 5 p.m., Monday to Friday, were normally billed at a day rate and calls between 5 p.m. and 11 p.m. on weekdays were usually billed at an evening rate, which reflected a discount of approximately 33 percent over the day rate. The last rate category, night and weekend, was applicable to calls made between 11 p.m. and 8 a.m. on weekdays as well as anytime on weekends and holidays. Calls during this rate period were usually discounted 50 percent from the day rate.

Table 3.1 contains a sample PSTN rate table that represented the cost of a PSTN called in the United States until the late 1990s and which resembles rate tables still in effect in some countries. This table is included for illustrative purposes and should not be used to determine the actual cost of a PSTN. Where this type of rate table is still in use, the cost of intrastate calls by state and interstate will more than likely vary. In addition, the cost of using different communications carriers to place a call between

Table 3.1 Sample PSTN Rate Table (Cost per Minute in Cents)

| | Rate Category | | | | | |
| Mileage between Locations | Day | | Evening | | Night and Weekend | |
	First Minute	Each Additional Minute	First Minute	Each Additional Minute	First Minute	Each Additional Minute
1–100	0.31	0.19	0.23	0.15	0.15	0.10
101–200	0.35	0.23	0.26	0.18	0.17	0.12
201–400	0.48	0.30	0.36	0.23	0.24	0.15

similar locations will typically vary from vendor to vendor. You should obtain a current state or interstate schedule from the vendor you plan to use to determine or project the cost of using PSTN facilities.

In the United States, most communications carriers initiated a flat rate structure during the late 1990s and early turn of the century. Flat rates for long-distance calls could be obtained for as low as five cents per minute, regardless of the time of day the call was made. To obtain this low rate, some carriers required customers to pay a monthly fee, which when added to the cumulative cost of calls could double the per minute fee for customers that made infrequent calls during the month. Other communications carriers introduced unlimited long-distance calling plans for one monthly fee regardless of a customer's call volume. Thus, in the United States subscribers have numerous options to consider.

3.2 LEASED VERSUS SWITCHED FACILITY ANALYSIS

Cost, speed of transmission, and degradation of transmission are the primary factors used in the selection process between leased and switched lines.

As an example of the economics associated with comparing the cost of PSTN and leased line usage, assume a personal computer located 50 miles from a mainframe has a requirement to communicate between 8 a.m. and 5 p.m. with the mainframe once each business day for a period of 30 minutes. Using the data in Table 3.1, each call would cost $0.31 \times 1 + 0.19 \times 29$ or $5.82. Assuming there are 22 working days each month, the monthly PSTN cost for communications between the personal computer (PC) and the mainframe would be 5.82×22 or $128.04. If the monthly cost of a leased line between the two locations was $450, it is obviously less expensive to use the PSTN for communications. Suppose the communications application lengthened in duration to 2 hours per day. Then, from Table 3.1, the cost per call would become $0.31 \times 1 + 0.19 \times 119$ or

Table 3.2 Line Selection Guide

Line Type	Distance between Transmission Points	Speed of Transmission	Use of Transmission
Dedicated (direct connect)	Local	Limited by conductor	Short or long duration
Switched (dial-up)	Limited by telephone access availability	Normally 56,000 bps	Short duration transmission
Leased (private)	Limited by communications carrier availability	Limited by type of facility	Long duration or many short duration calls

$22.92. Again assuming 22 working days per month, the monthly PSTN charge would increase to $504.24, making the leased line more economical.

If your organization benefits from the use of a flat rate plan, the cost computations are relatively simple. For example, assume your organization pays 10 cents per minute for PSTN calls, regardless of the time they are made or the distance to the called party. If you have a computer that requires 30 minutes of communications per day to the mainframe, the dial cost becomes 30 minutes/day × 22 work days/month × 10 cents/minute or $66 per month.

As a rule of thumb, if data communications requirements to a computer involve occasional random contact from a number of terminals at different locations, and each call is of short duration, dial-up service is normally employed. If a large amount of transmission occurs between a computer and a few terminals, leased lines are usually installed between the terminal and the computer.

Because a leased line is fixed as to its routing, it can be conditioned to reduce errors in transmission as well as permit ease in determining the location of error conditions because its routing is known. Normally, switched circuits are used for transmission at speeds up to 56,000 bps. Some of the limiting factors involved in determining the type of line to use for transmission between terminal devices and computers are listed in Table 3.2.

3.2.1 Servicing Additional Terminal Devices

Now let us assume that instead of 1 terminal device there are 4, each requiring 30 minutes of access per day to the distant mainframe. As previously computed using Table 3.1, the monthly cost of using the PSTN would be $128.04 per terminal or $512.16 for 4 terminals, each having a daily transmission requirement of 30 minutes. However, if each terminal device were to use a separate line the relationship between the total switched network cost and leased line cost would remain the same, negating any rationale for comparing the cost of the transmission cost of individual terminal devices using the PSTN to the cost of providing communications via multiple leased lines. Fortunately, communications equipment manufacturers have developed a variety of products that can be employed to share the use of a leased line among terminal devices. These products include a variety of routers, multiplexers, data concentrators, and line sharing devices whose operation and use are described in subsequent chapters of this book.

3.2.2   Line Sharing Economics

To illustrate the potential advantages associated with the use of line sharing devices, let us examine how a multiplexer or router could be used to share the use of a leased line among many terminal devices and the economics associated with the use of such equipment.

Figure 3.2 illustrates two methods by which a pair of routers or multiplexers could be used to enable four terminal devices to share the use of a common leased line. In the top portion of Figure 3.2, it is assumed that each terminal device is within close proximity of the multiplexer or router, enabling each device to be directly cabled to the communications device located at the remote site. In the lower portion of Figure 3.2, a diametrically opposite situation is assumed. Here, the terminal devices are assumed to be dispersed within the same or different buildings and accessing the multiplexer or router is accomplished via local PSTN calls. Not shown, but perhaps obvious, are a variety of configurations representing combinations of directly connected terminals and terminals accessing the multiplexer or router via the PSTN.

In the wonderful world of data communications, the modems (M) shown in Figure 3.2B are employed for use over the PSTN, while the service units in Figure 3.2A and Figure 3.2B imply digital leased lines are used. In the United States, essentially all backbone circuits connecting major population centers were converted to digital many years ago. However, in certain areas of the globe analog circuits continue to form the backbone of some carriers. When this occurs, the service units would be replaced by high-speed modems.

Now that we have reviewed the basic use of routers and multiplexers to share the transmission capability of a leased line among a number of data sources, let us focus our attention upon the economics of their use. This will allow us to observe that, in many instances, the cost of a leased line may represent only a small portion of the cost of using a leased line.

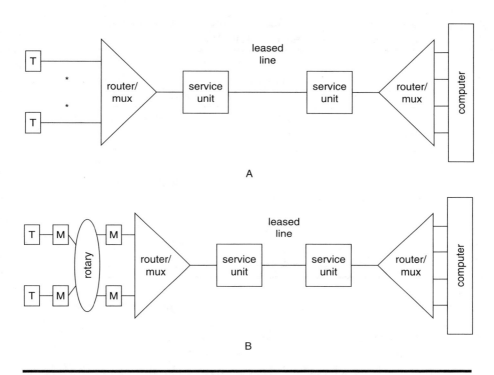

Figure 3.2   Using Routers or Multiplexers to Share a Leased Line
(A) Directly connecting terminal devices. (B) Using the PSTN to access a remote router or multiplexer.

Because the cost of cabling is a onetime charge and leased lines are billed monthly, without considering whether or not you will purchase or lease service units and routers or multiplexers, you will still have to determine how to proportion onetime costs into a monthly basis to be able to obtain a comparison to monthly PSTN usage. To do so, most organizations will first attempt to determine a realistic system life for any hardware purchased and then divide the system life in months into any onetime costs to obtain an equivalent monthly cost. Although this method ignores the cost of money when making an initial onetime expenditure, it represents a fairly accurate method during periods of low to moderate interest rates for a system life under four years. Because the system life of most communications equipment is typically three years or less, we will use this method and ignore the cost of money in the economic analysis performed in this chapter.

In comparing the cost of the network configuration illustrated in Figure 3.2A to the use of the PSTN shown in Figure 3.2B, let us make several cost assumptions. First, let us assume that four terminal devices, such as

Table 3.3 System Life and Monthly Cost Computation for Directly Connected Terminal Devices

	Onetime Cost ($)	Monthly Recurring Cost ($)
Terminal cabling (4 @ $25)	100	
Routers/multiplexers (2 @ $1000)	2000	
Service units (2 @ $500)	1000	
Leased line	1500	450
	4600	450
System life cost computation		
Onetime	4600	
Recurring ($450 × 36 months)	16,200	
System life cost	20,800	
Monthly cost ($20,800/36 months)		578

personal computers, will be cabled directly to the router or multiplexer at the remote site and the cables cost $25 per terminal. Next, assume that the routers or multiplexers cost $1000 per unit and each service unit costs $500 and the leased line has a $1500 installation charge and a $450 monthly recurring cost. If we assume a 36-month system life, the cost of the directly connected system illustrated in Figure 3.2A as computed in Table 3.3 is $20,800. Dividing that cost by the 36-month system life results in a monthly cost of $578.

Prior to computing the system cost associated with using the PSTN to access the remote router or multiplexer illustrated in Figure 3.2B, let us note the differences between that network configuration and the network configuration illustrated in Figure 3.2A. In Figure 3.2B, the use of the PSTN requires each terminal device to have an access line to dial a modem connected to the remote router or multiplexer. Similarly, each dial-in line at the multiplexer also requires a PSTN access line. Thus, if no contention is designed into the system, 2 × T access lines will be required, where T equals the number of terminal devices to be serviced. This is because when terminal devices communicate over the PSTN, modems will be required for each terminal and each dial-in line as illustrated in Figure 3.2B.

In analyzing the economics associated with the configuration illustrated in Figure 3.2B, the cabling costs associated with Figure 3.2A have been replaced by the cost of eight access lines and eight modems. Otherwise, the cost of the routers or multiplexers, service units, and leased lines remains the same. Assuming the installation cost of each access line is $50 with a monthly cost of $25 and the cost of the modems connected

Table 3.4 System Life and Monthly Cost Computation for Using the PSTN to Access the Router or Multiplexer

	Onetime Cost ($)	Monthly Recurring Cost ($)
Terminal cabling (4 @ $25)	100	
Access lines	400	200
Modems (8 @ $100)	800	
Routers/multiplexers (2 @ $1000)	2000	
Service units (2 @ $500)	1000	
Leased line	1500	450
	5700	650
System Life Cost Computation		
Onetime	5700	
Recurring ($650 × 36 months)	23,400	
System life cost	28,800	
Monthly cost ($28,800/36 months)		800

to the terminals and dial-in lines at the multiplexer is $100 per unit, Table 3.4 analyzes the system life and monthly cost of the PSTN multiplexer configuration. As indicated in Table 3.4, the monthly cost of a multiplexed leased line system using the PSTN to access the remote multiplexer is $800 when onetime costs are amortized over the 36-month system life.

A comparison of the monthly cost of the configurations illustrated in Figure 3.2A and Figure 3.2B illustrates an important communications concept. Whenever possible, it is normally economically advantageous to bypass the PSTN when concentrating data sources that can be directly cabled to a device. Of course, when terminals are widely dispersed within a geographical area, you cannot directly cable those devices to a router or multiplexer. However, as the number of terminal devices to be serviced increases, it becomes possible to economize upon costs by limiting the number of dial-in access lines and modems at the remote router or multiplexer. This process is known as sizing and is explained in detail in Chapter 6.

Until now, we have simply computed the PSTN cost without considering the total cost associated with using that facility to access a distant computer. Now let us expand upon the PSTN cost by treating that access method as a system and examine its various cost components.

Figure 3.3 illustrates the typical method by which terminal devices would use the PSTN to access remote computational facilities. Note that this configuration is similar to the remote site configuration previously

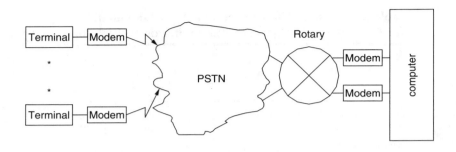

Figure 3.3 PSTN Access to Remote Computer

Table 3.5 AT&T Multirate Schedule Tariff FCC No.1, Effective November 29, 1989, Direct-Dialed Station-to-Station Usage

Rate Mileage	Day Initial Minute ($)	Day Additional Minute ($)	Evening Initial Minute ($)	Evening Additional Minute ($)	Night and Weekend Initial Minute ($)	Night and Weekend Additional Minute ($)
1–10	0.1800	0.1700	0.1206	0.1139	0.1000	0.0975
11–22	0.2100	0.2000	0.1407	0.1340	0.1130	0.1100
23–55	0.2300	0.2200	0.1541	0.1474	0.1200	0.1200
56–124	0.2300	0.2200	0.1541	0.1474	0.1200	0.1200
125–292	0.2300	0.2300	0.1541	0.1541	0.1215	0.1215
293–430	0.2400	0.2300	0.1608	0.1541	0.1250	0.1225
431–925	0.2400	0.2400	0.1608	0.1608	0.1300	0.1260
926–1910	0.2500	0.2500	0.1675	0.1675	0.1325	0.1300
1911–3000	0.2500	0.2500	0.1675	0.1675	0.1350	0.1325
3001–4250	0.3100	0.3000	0.2077	0.2010	0.1650	0.1600
4251–5750	0.3300	0.3200	0.2211	0.2144	0.1750	0.1700

illustrated in Figure 3.2B, with the multiplexer replaced by the computer while the requirement for a pair of high-speed modems is eliminated because no leased line is required.

Assuming that long-distance calls will be required to access the computer from each terminal location, you will want to examine the rate schedule of several communications carriers, such as AT&T, Verizon, and Sprint. Table 3.5 contains the rate schedule for AT&T long-distance service that was in effect in early 1990 and which we will use to determine the cost of long-distance calls in the following analysis. Note that although Table 3.5 is obviously obsolete, its use illustrates an economic analysis

Table 3.6 PSTN System Life and Monthly Cost Computation

| | Onetime Cost ($) | Monthly Recurring Cost | | |
		10 min/day ($)	30 min/day ($)	60 min/day ($)
Access lines	400	200	200	200
Long-distance calls		202	607	1214
Modems (8 @ $100)	800			
	1200	402	807	1414
System life cost computation				
Onetime		1200	1200	1200
Recurring (monthly × 36)		14,472	29,052	50,904
System life cost		15,672	30,252	52,104
Monthly cost (system life/36)		435	840	1447

performed with a complex rate schedule that in many areas of the world is still in use. In fact, although most communications carriers in the United States now use a flat rate schedule and heavily discount that schedule for large use organizations, some carriers have discussed a return to a multirate schedule. Returning to a multirate schedule would encourage additional usage during evenings and weekends and more accurately recover the costs associated with operating their network.

Table 3.6 indicates the PSTN system life and monthly cost comparison based upon the assumption that each terminal user requires an average of 10, 30, and 60 minutes of daily transmission to the remote computer 22 days per month. Assuming there is 1 call per day and the terminals are located 200 miles from the computer center, according to Table 3.5 the initial minute and each additional minute cost $0.23. Thus, 1 hour of transmission would cost $13.80 and the monthly cost for 1 hour of transmission per day would be $13.80/hour × 1 hour/day × 22 days/month or $303.60. Note that the cost computations in Table 3.6 assume that access lines are required for each terminal as well as 4 access lines at the computer site. This results in a onetime cost of $400 and a monthly recurring cost of $200 based upon our previous assumption that the installation cost of an access line is $50 with a monthly cost of $25.

Based upon the previous assumptions, the use of the PSTN 10 minutes per day is more economical than using routers or multiplexers for data concentration, regardless of the method terminals employ to access the data concentration system. As the average duration of PSTN usage increases, the cost of PSTN usage increases. This results in an average PSTN daily usage of 60 minutes per day exceeding the cost of both methods of accessing a router or multiplexer data concentration system.

Note that the monthly cost difference between PSTN usage for 60 minutes per day ($1447) and the use of a router or multiplexer where access is gained via the PSTN ($800) is significant. However, at a daily usage of 60 minutes per day, the cost of using the PSTN ($840) slightly exceeds the cost of using data concentration equipment ($800). In this situation, you might consider continuing to use the PSTN to determine an average cost over a period of months prior to purchasing equipment necessary to install a router or multiplexer system and become committed to a leased line. However, if you observed or estimated an average of 60 minutes of PSTN usage per terminal device, the cost differences would be of a sufficient magnitude for most organizations to justify the use of a router or multiplexer system regardless of whether or not there could be some small variance in the average terminal daily transmission figure.

3.2.3　Concepts to Consider

A review of the PSTN monthly costs based upon an average transmission of 10, 30, and 60 minutes per day illustrates several important concepts. First, even a minimal PSTN usage can have a substantial monthly cost due to the repeating cost of access lines and the amortization of onetime costs, such as the installation of access lines and the cost of modems. Second, the cost of PSTN usage is not actually uniform. As indicated in our example, the cost for 30 minutes per day of transmission is a little more than twice the cost of using the PSTN 10 minutes per day, while the cost for an average transmission of 60 minutes per day is approximately three times the cost associated with an average transmission of 10 minutes per day. The primary reasons for the nonuniform cost relationship include the onetime costs that are not based upon usage and the monthly access line fees that are also billed regardless of PSTN usage. These are the facts regardless of whether a multirate schedule similar to Table 3.5 or a uniform single rate schedule is associated with the cost of using the PSTN.

Although actual economic computations are no substitute for generalizations, we can summarize several economic concepts with respect to the use of the PSTN versus the use of leased lines. First, as the number of terminal devices accessing a computer facility and their transmission duration increase, their cumulative monthly PSTN costs will increase. This is illustrated by Curve C in Figure 3.4, which provides a general comparison of the costs of using the PSTN as opposed to the use of routers or multiplexers to aggregate multiple data sources onto a leased line. When transmission time is minimal, the cost associated with the use of the PSTN is minimized; however, there is always an access charge and there may be onetime charges that form a base to which the cost of the calls builds upon. Second, as the duration of communications increase, long-distance PSTN charges will proportionally increase. Thus, both the number of data

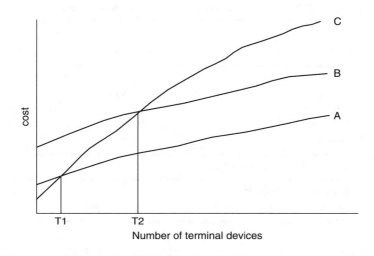

Figure 3.4 PSTN Plotted against Multiple Leased Lines
Curve A — data concentration with directly cabled terminals. Curve B — PSTN
access to data concentrator. Curve C — PSTN access to remote computer.

sources and the duration of their transmission will govern the majority of
the costs associated with the use of PSTN facilities as both the number
of data sources and their transmission duration increase.

With regard to the use of routers, multiplexers, or other types of data
concentration devices located at a remote site, we previously noted that
direct cabling will be less expensive than PSTN access to a data concentration device. Curve A in Figure 3.4 represents the cost of data concentration when data sources are directly cabled to the data concentration
device. This curve is relatively flat, with the increase in cost associated
with an increasing number of terminals serviced by the data concentrator
equipment resulting from the additional onetime cabling cost for each
terminal. Of course, a point will be reached where the router, multiplexer
or similar device, and high-speed modems or service units connected to
analog or digital leased lines may have to be replaced by devices with a
higher capacity. The cost to expand routers or multiplexers or to acquire
higher speed modems or service units is not shown in Figure 3.4. The
cost for the expansion or acquisition would raise the curve at T_n, where
T_n indicates the number of terminals that require either servicing by large
capacity routers or multiplexers or the use of higher speed modems or
service units or possibly both.

Curve B in Figure 3.4 represents the generalized cost of providing
PSTN access to a data concentration device. Note that this curve is always
higher than Curve A and will similarly jump to a new height at T_n.

Table 3.7 AT&T Analog Leased Line Rate Schedule

| Interoffice Channel | Monthly Charge | |
Mileage Band	Fixed ($)	Per Mile ($)
1–50	72.98	2.84
51–100	149.38	1.31
101–500	229.28	0.51
501–1000	324.24	0.32
1000+	324.24	0.32

Central Office Connection	
Monthly charge ($)	16.40
Nonrecurring charge ($)	196

Two other general observations are warranted concerning Figure 3.4, which are the break-even points labeled T_1 and T_2. T_1 represents the number of terminal devices whose total switched network system cost equals the cost of routing, multiplexing, or data concentration when terminals are directly cabled to a data concentration device. In general, this break-even point occurs far quicker than T_2, which represents the break-even point between using the PSTN and using a data concentration device where access to the device is via the PSTN.

Until now, we have simply assumed a fixed monthly cost for a leased line. To indicate how the cost of a leased line varies by mileage, AT&T's analog private line rate schedule that was in effect in early 1990 is reproduced in Table 3.7. Note that the mileage band charges contained in Table 3.7 only apply to the mileage between the central offices serving each customer location. As such, they do not include the cost of local access from a subscriber's premises to an AT&T central office where the AT&T point of presence is located. Note that the central office monthly connection charge is per central office and the nonrecurring charge represents the onetime cost for connecting the local access line to AT&T's central office. Later in this chapter, we will examine the cost of certain types of digital transmission facilities to expand upon our use of rate tables.

To illustrate the computation of interoffice channel charges, assume the mileage between offices was 1832. From Table 3.7, the fixed monthly charge would be $324.24 and the monthly charge per mile would be $0.32. Thus, the monthly cost of a 1832-mile interoffice channel would be $324.24 + 1832 miles × $0.32/mile or $910.48.

3.3 ADDITIONAL ANALOG FACILITIES TO CONSIDER

There are several additional types of analog facilities offered by communications carriers to consider in the network design process. Two of the most common facilities are WATS and FX. Each of these facilities has its own set of characteristics and rate structure and requires an analytic study to determine which type or types of service should be used to provide cost-effective service for the user. In this section, we will examine the operation and utilization of WATS and FX facilities and compare and contrast their use to PSTN and leased line facilities. In doing so, we will assume certain costs associated with each service for comparison purposes; however, you should contact your communications carrier to determine the actual cost of each service you are considering.

3.3.1 WATS

Introduced by AT&T for interstate use in 1961, WATS is now offered by most communications carriers. Its scope of coverage has been extended from the continental United States to Hawaii, Alaska, Puerto Rico, the U.S. Virgin Islands, and Europe, as well as selected Pacific and Asian countries.

WATS may be obtained in two different forms, each designed for a particular communications requirement:

1. Outward WATS — used when a specific location requires placing a large number of outgoing calls to geographically distributed locations.
2. Inward WATS — permits a number of geographically distributed locations to communicate with a common facility.

Calls on WATS are initiated in the same manner as a call placed on the PSTN. However, instead of being charged on an individual call basis, the user of WATS facilities normally pays a flat rate per block of communications hours per month occurring during weekday, evening, and night and weekend time periods. Over the past few years, AT&T and other communications carriers have introduced a variety of WATS plans that small businesses, as well as large volume users, can consider. Some of these WATS plans bill outward calls based upon the mileage to the called location, the time of day, and an initial 30-second period, as well as 6-second increments. Other WATS plans provide customers with a block of hours for calls to a particular state or regional area for a fixed monthly fee. For illustrative purposes, let us turn our attention to Table 3.8, which contains the AT&T PRO WATS Tariff No. 1 that was effective in early 1990.

Table 3.8   AT&T PRO WATS Tariff Intramainland and Mainland to Hawaii and Alaska — Dial Station Rates

Rate Mileage	Initial 30 s or Fraction			Each Additional 6 s or Fraction		
	Day ($)	Evening ($)	Night ($)	Day ($)	Evening ($)	Night ($)
0–55	0.0960	0.0655	0.0485	0.0192	0.0131	0.0097
56–292	0.1080	0.0740	0.0565	0.0216	0.0148	0.0113
293–430	0.1155	0.0795	0.0600	0.0231	0.0159	0.0120
431–925	0.1195	0.0825	0.0640	0.0239	0.0165	0.0128
926–1910	0.1245	0.0855	0.0665	0.0249	0.0171	0.0133
1911–3000	0.1245	0.0855	0.0665	0.0249	0.0171	0.0133
3001–4250	0.1505	0.1035	0.0805	0.0301	0.0207	0.0161
4251–5750	0.1605	0.1105	0.0860	0.0321	0.0221	0.0172

Although the tariff has been adjusted several times and organizations with significant use of a mixture of AT&T facilities can expect volume discounting, the referenced table provides a general indication of the manner by which WATS usage is billed.

In addition to the tariffs listed in Table 3.8, you must consider the cost of the WATS access line as well as volume discounts provided based upon usage volume. In general, PRO WATS provides at least a 10 percent discount in comparison to the use of the PSTN and organizations with large out-dial requirements may easily experience a 15 percent or greater discount over the cost of using the switched network.

A voice-band trunk called an access line is provided to the WATS users. This line links the facility to a telephone company central office. Other than cost considerations and certain geographical calling restrictions that are a function of the service area of the WATS line, the user may place as many calls as desired on this trunk if the service is outward WATS or receive as many calls as desired if the service is inward.

For data communications users, outward WATS, such as AT&T's PRO WATS, is normally used to provide communications between a central computer facility and geographically dispersed terminal devices that can be polled. In this type of communications environment, the central computer will use an auto-dial modem during the evening or nighttime, when rates are low, to poll each terminal device. Left powered on, the terminal devices will have an auto-answer modem and respond to the poll by transmitting previously batched data traffic to the computer for processing. Once all remote locations are polled and any required processing is completed, the central computer may again dial each terminal, transmitting the processed data to its appropriate destination.

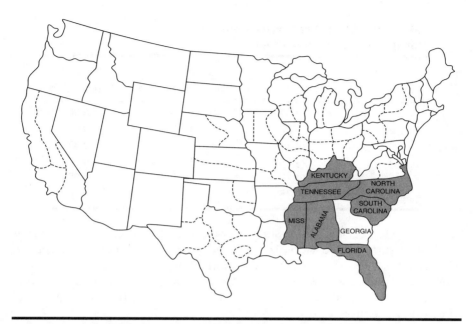

Figure 3.5 AT&T WATS Service Area 1 for an Access Line Located in Georgia

Because access lines and modems would be required for PSTN communications, the only difference in cost between outward WATS and the use of the PSTN involves the higher access line costs associated with WATS and its usage costs, which are usually 10 to 15 percent below the cost of using the PSTN. In general, PSTN long-distance usage exceeding $150 per month is the break-even point between outward WATS and PSTN usage, with WATS providing a 10 to 15 percent savings over the PSTN as the monthly cost of PSTN usage increases over $150.

Inward WATS, such as the well-known 800 area code, permits remotely located personnel to call your facility toll-free from the service area provided by the particular inward WATS-type of service selected. The charge for WATS is a function of the service area. This can be intrastate WATS, a group of states bordering the user's state where the user's main facility is located, a grouping of distant states, or International WATS, which extends inbound 800 service to the United States from selected overseas locations. Another service similar to WATS is AT&T's 800 READYLINE® service. This service is essentially similar to WATS, however, calls can originate or be directed to an existing telephone in place of the access line required for WATS service.

Figure 3.5 illustrates the AT&T WATS Service Area 1 for the state of Georgia. If this service area is selected and a user in Georgia requires

Table 3.9 AT&T 800 Service Rate Schedule (for Illustrative Purposes Only)

Service Area	Per Hour of Use		
	Business Day ($)	Night/Evening ($)	Weekend ($)
1	13.42	11.06	8.90
2	13.88	11.43	8.92
3	14.11	11.61	9.35
4	14.55	11.99	9.65
5	14.79	12.18	9.80
6	16.15	13.30	10.70

inward WATS service, he or she will pay for toll-free calls originating in the states surrounding Georgia — Florida, Alabama, Mississippi, Tennessee, Kentucky, South Carolina, and North Carolina. Similarly, if outward WATS service is selected for Service Area 1, a person in Georgia connected to the WATS access line will be able to dial all telephones in the states previously mentioned.

The states comprising a service area vary based upon the state in which the WATS access line is installed. Thus, the states in Service Area 1 when an access line is in New York state would obviously differ from the states in a WATS Service Area 1 when the access line is in Georgia. Fortunately, AT&T publishes a comprehensive book that includes maps of the United States, illustrating the composition of the service areas for each state. AT&T also publishes a time-of-day rate schedule for each state based upon state service area.

To illustrate the cost of inward WATS, we will examine the cost elements associated with its use, referencing a common type of inward WATS rate schedule. Table 3.9 contains an example of an AT&T 800 service rate schedule, which is shown only for illustrative purposes because rates frequently change. The service areas numbered one to six refer to increasing bands of states surrounding the state the user is located in. Neither area installation and access line charges, nor a service group charge and usage discounts that vary from 5 percent for charges exceeding $50 in a month to 15 percent for charges exceeding $1350 in a month are shown. Concerning the service group charge, its fee of $20 per month per 800 number or band grouping is for the rotary or hunt group that automatically transfers calls to the next access line if the first access line is busy.

For cost comparison purposes, assume that we want to service four geographically distributed terminal devices that will gain access to a computer system via the use of 800 service. Let us commence our analysis using four access lines and one hour per day of transmission similar to our previous PSTN economic analysis.

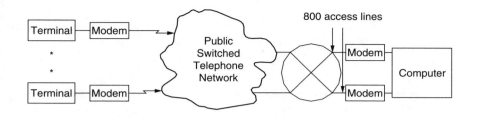

Figure 3.6   Using 800 Service for Access to a Central Computer Site

Table 3.10   800 Service System Cost Computation

	Onetime Cost ($)	Monthly Recurring Cost ($)
Terminal access lines	200	100
WATS access lines	267	148
Service group		20
Low-speed modems (8 @ $100)	800	
WATS usage 22 hrs/mo × 4 × 13.42[a]		1098
	1267	1366
System Life Cost Computation		
Onetime	1267	
Recurring ($1366 × 36 months)	49,176	
System life cost	50,443	
Monthly cost (50,443/36)		1401

[a] Amounts in excess of $350, up to $1350, are reduced by 10 percent.

Figure 3.6 illustrates the use of 800 service to satisfy the communications requirements of four terminal users. Let us also assume that each terminal will be serviced with a separate WATS access line, ensuring the elimination of contention. Using a WATS access line charge of $267 and a recurring access line fee of $37 per month we can compute the cost for four terminal devices located in Service Area 1 outside of Georgia using 800 service to access a computer in that state. Table 3.10 indicates the system life and monthly cost computation to service the previously described transmission requirements via inward WATS. As indicated in Table 3.10, the expected monthly cost would be $1366, which is slightly less than the cost of using the PSTN one hour per day as computed in Table 3.6. Although the break-even transmission time between PSTN and 800 service will vary based upon access line charges, PSTN and WATS tariffs in effect, and average daily transmission duration, in general, savings from 800 service begin to accrue when daily transmission exceeds approximately

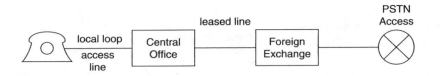

Figure 3.7 FX Line Components
The FX line combines the use of a leased line and switched network access. When the local telephone is placed in an off-hook condition, a dial tone from the FX is received.

one hour per day. Thereafter, as transmission increases beyond one hour per day, 800 service will provide additional savings over the use of the PSTN.

3.3.2 Foreign Exchange

FX service may provide a method of transmission from a group of terminal devices remotely located from a central computer facility at less than the cost of direct distance dialing. An FX line can be viewed as a hybrid combination of a leased line and PSTN access.

Figure 3.7 illustrates the major components of an FX line and their relationship to one another. The access line is the local loop routed from the subscriber's premises to the serving telephone company's central office. The leased line connects the central office serving the end user to a distant central office at which a connection to the PSTN occurs. That distant office is known as the foreign exchange, hence the name used to describe this hybrid combination of leased line and PSTN access line.

An FX line can be used for both call origination and call reception. When the local telephone is placed off-hook, the subscriber receives a dial tone from the foreign exchange. Then, long-distance calls to the foreign exchange area can be made as local calls. However, the total cost of FX service will include the monthly cost of the access line to the central office, the leased line between the central office and the foreign exchange, as well as a monthly charge from the operating company providing the interconnection to the PSTN at the foreign exchange. The last charge can range from a flat monthly fee to a flat fee plus a usage charge, typically ranging between $0.05 and $0.10 per minute of use.

When used for call reception, an FX line permits end users to dial a local number to obtain long-distance toll-free communications. Figure 3.8 illustrates how an FX line could be used to service a number of terminal users located at a remote site. In this example, each terminal user would dial a local number associated with the foreign exchange and obtain a long-distance connection to the central site via a local call.

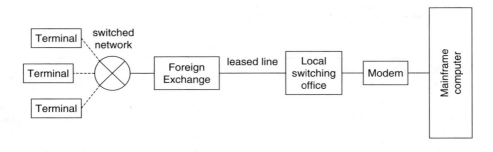

Figure 3.8 FX Service
An FX line permits many terminal devices to use the facility on a scheduled or contention basis.

The use of an FX line eliminates long-distance charges that would be incurred by users directly dialing the distant computer facility. Because only one person at a time may use the FX line, normally only groups of users whose usage can be scheduled are suitable for FX utilization — if you wish to install a single FX line. Otherwise, you can consider installing additional FX lines to reduce or eliminate potential contention between a number of remote terminal users and the number of FX lines available for connection to the central site computer.

The major difference between an FX line and a leased line is that any terminal device dialing the FX line provides the second modem required for the transmission of data over the line, whereas a leased line used for data transmission normally has a fixed modem or service unit attached at both ends of the circuit.

3.3.2.1 Economics of Use

The primary use of an FX line is to provide toll-free calls to or from a relatively small geographical area serviced by a foreign exchange. As such, its economics of utilization should be compared to the use of the PSTN and WATS.

To understand the economics associated with the use of FX lines, let us make four assumptions concerning the cost of the various components of this type of circuit:

1. The access line billed by the local operating company has an installation cost of $50 and a monthly recurring cost of $25.
2. The leased line portion of the FX line has an installation cost of $200 and a monthly recurring cost of $450.

Table 3.11   FX Line System Life Cost Computation

	Onetime Cost ($)	Monthly Recurring Cost ($)
Closed end access line	50	25
Leased line	200	450
Open end access line	50	25
Usage (60 × 22 × 0.08)		106
Modems (2 @ $100)	200	
	500	606
System Life Cost Computation		
Onetime	500	
Recurring ($606 × 36)	21,816	
System life cost	22,316	
Monthly cost (22,316/36)		620

3. The open end of the FX line maintained by a distant operating company has an installation cost of $50, a monthly recurring cost of $25, and a usage cost of $0.08 per minute.

4. The modem used by a terminal user and the modem connected to the closed end of the FX line each cost $100 and the system life of the line will be 36 months.

Based upon the preceding assumptions, Table 3.11 summarizes the system life computations for an average transmission of 1 hour per day, 22 business days per month.

An analysis of the costs associated with the use of the FX line contained in Table 3.11 illustrates several important concepts concerning the economics associated with the use of a foreign exchange line. First, the recurring costs over the system life will normally substantially exceed the onetime cost to establish this type of communications system. Second, for a low volume of transmission, the cost per minute of an FX line will substantially exceed the cost of using the PSTN or WATS. This can be seen by dividing the monthly recurring cost of $606 by the 1320 minutes of transmission per month (60 × 22), which results in a per minute recurring cost of approximately $0.46. Only when an FX line carries a substantial volume of transmission will its cost per minute decrease to where it will be cost-effective with the PSTN and WATS. As an example of this, consider the economics associated with increasing the transmission over the FX line to a daily average of 5 hours or 300 minutes. If this occurs, the monthly recurring cost would increase to $1028, resulting in a decrease in the per minute cost to approximately $0.16 per minute per month.

Because the per-minute cost of an FX line will normally exceed the cost of using the PSTN or WATS, it is reasonable to ask, "Why use this line facility for data transmission?" In comparison to PSTN usage, FX allows toll-free access. Thus, an organization installing an FX line could provide the access number for its customers to use without requiring the customer to make long-distance calls or collect calls that could not be accepted by a computer connected to a PSTN access line. In comparison to WATS, an FX line can be used for both incoming and outgoing calls, whereas some WATS lines are restricted to either incoming or outgoing calls. Thus, a group of FX lines under certain circumstances may provide a more efficient data transmission mechanism than the use of groups of inward and outward WATS lines.

3.4 DIGITAL FACILITIES

In addition to analog service, communications carriers have implemented numerous digital service offerings over the last two decades. Using a digital service, data is transmitted from source to destination in digital form without the necessity of converting the signal into an analog form for transmission over analog facilities, as is the case when modems are used for data transmission. Although all communications carriers provide some type of digital service it is beyond the scope of this book to discuss each service. Thus, we will focus our attention upon representative services by describing AT&T and British Telecom offerings, which other carriers market under various trade names.

3.4.1 AT&T Offerings

In the United States, AT&T offers several digital transmission facilities under the Accunet® Digital Service. Dataphone Digital Service (DDS) was the charter member of the Accunet family and is deployed in over 100 major metropolitan cities in the United States, as well as having an interconnection to Canada's digital network. DDS operates at synchronous data transfer rates of 2.4, 4.8, 9.6, 19.2, and 56 kbps, providing users of this service with dedicated, two-way simultaneous transmission capability.

Originally, all AT&T digital offerings were leased line services where a digital leased line is similar to a leased analog line in that it is dedicated for full-time use to one customer. In the late 1980s, AT&T introduced its Accunet Switched 56 service, a dial-up 56 kbps digital data transmission service. This service enables users to maintain a dial-up backup for previously installed 56 kbps DDS leased lines as well as other types of higher speed digital transmission facilities. In addition, this service can be used to supplement existing services during peak transmission periods or for applications that only require a minimal amount of transmission time per day, because the service is billed on a per minute basis.

Two other offerings from AT&T are Accunet T1.5 Service and Accunet Spectrum of Digital Services (ASDS). Accunet T1.5 Service is a high-capacity 1.544 Mbps terrestrial digital service that permits 24 voice-grade channels or a mixture of voice and data or just high-speed data to be transmitted in digital form. This service was originally only obtainable as a leased line and is more commonly known as a T1 channel or circuit. During the late 1990s, AT&T and other communication carriers expanded their dial-up digital network offerings, permitting customers in certain locations to be able to subscribe to a dial T1 transmission service.

ASDS provides end-to-end single channel digital transmission at data rates ranging from 56 or 64 kbps up to 768 kbps in 64 kbps increments. Due to this transmission structure, ASDS is also commonly referred to as FT1 service.

Other offerings from AT&T are primarily oriented toward large organizations, such as T3 service, which provides the equivalent of 28 T1 lines and operates at 44.736 Mbps and several Synchronous Optical Network (SONET) facilities that support data rates in the hundreds of millions of bits per second. Because this book is oriented toward a majority of, but not all, persons who work in the field of data communications, we elected to cover the more popular digital offerings instead of those offerings that are only used by the largest commercial organizations and governmental agencies.

3.4.2 European Offerings

In Europe, most countries have established digital transmission facilities. One example of such offerings is British Telecom's KiloStream service. KiloStream provides synchronous data transmission at 2.4, 4.8, 9.6, 48, and 64 kbps and is similar to DDS. Each KiloStream circuit is terminated by British Telecom with a network terminating unit (NTU), which is the digital equivalent of the modem required on an analog circuit and is better known as a DSU in North America.

3.4.2.1 Data Service Units

A DSU provides a standard interface to a digital transmission service and handles such functions as signal translation, regeneration, reformatting, and timing. The DSU is designed to operate at one of five speeds — 2.4, 4.8, 9.6, 19.2, and 56 kbps. The transmitting portion of the DSU processes the customer's signal into bipolar pulses suitable for transmission over the digital facility. The receiving portion of the DSU is used both to extract timing information and to regenerate mark and space information from

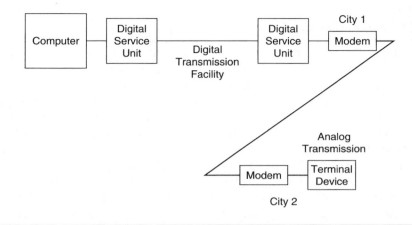

Figure 3.9   Analog Extension to Digital Service
Although data is transmitted in digital form from the computer to City 1, it must be modulated by the modem at that location for transmission over the analog extension.

the received bipolar signal. A second interface arrangement for DDS is a CSU and at one time was provided by the communications carrier to those customers who wished to perform the signal processing to and from the bipolar line, as well as to retime and regenerate the incoming line signals using their own equipment. Today, most carriers as well as third party manufacturers market combined DSU/CSU devices, which we will refer to as simply DSUs.

As data is transmitted over digital facilities, the communications carrier regenerates the signal numerous times prior to its arrival at its destination. In general, digital service gives data communications users improved performance and reliability when compared to analog service, owing to the nature of digital transmission and the design of digital networks. This improved performance and reliability because digital signals are regenerated, whereas when analog signals are amplified, any distortion to the analog signal is also amplified.

Although digital service is offered in many locations, for those locations outside the serving area of a digital facility, the user will have to employ analog devices as an extension to interface to the digital facility. The use of digital service via an analog extension is illustrated in Figure 3.9. As depicted in Figure 3.9, if the closest city to the terminal located in City 3 that offers digital service is City 1, then to use digital service to communicate with the computer an analog extension must be installed between the terminal location in City 2 and City 1. In such cases, the performance, reliability, and possible cost advantages of using digital service may be completely dissipated.

3.4.3 Economics of Use

Until late 1989, users considering digital transmission would primarily compare the potential use of DDS and Accunet T1.5 to single and multiple analog leased lines. In late 1989, AT&T and other communications carriers began to introduce FT1 facilities, which resulted in users considering the economics associated with the use of this transmission facility with DDS and Accunet T1.5. In this section, we will examine examples of the structure of rate schedules associated with each of the three digital facilities that will form a basis for understanding the rationale for selecting an appropriate digital transmission facility. In addition, we will compare and contrast the rate structure of DDS to that of an analog line to understand why the use of DDS, in many instances, may not represent the least economical method to satisfy one's transmission requirements.

3.4.4 Dataphone Digital Service

Similar to the cost of an analog leased line, the cost of a DDS line is primarily based upon the distance in miles between AT&T central offices. The top portion of Table 3.12 indicates the DDS rate schedule in effect in early 1990 and the lower portion of the table shows the rates for the same service under Tariff FCC No. 9, effective April 20, 2001. The rationale for showing the evolution of DDS rates provides an opportunity to note a common misconception where many persons believe that digital transmission rates always decrease. If you compare the rates in Table 3.12A to that shown in Table 3.12B, you will note that at each data rate, the newer tariff implies a higher charge. However, to be fair to AT&T, it should be noted that DDS is considered by many persons to represent an obsolete digital offering that was introduced over 30 years ago and one that the carrier is more than likely attempting to migrate users off of and onto their newer digital offerings.

In addition to the mileage cost indicated in Table 3.12, there are several other costs to consider. These costs include the access lines to AT&T POP from each customer location, a flat rate AT&T bills customers for their central office DDS connection, and the cost of DSUs. To illustrate the cost of DDS service, let us assume you are are considering the installation of a 450-mile 56 kbps digital circuit. Assume the local access lines and DDS central office connections have a onetime cost of $250 and a monthly recurring cost of $100. Table 3.13 indicates the computations required to determine the system life cost for the use of 56 kbps service under the tariff in effect as of 2001, which is shown in Table 3.12B. Note that the monthly DDS mileage cost is determined by adding the fixed cost of $1111.00 to $3.50 per mile times 450 miles, because the 450 miles of the DDS circuit is in the 101 to 500 mileage band and the per-mile cost is based upon total circuit length falling in that band.

Table 3.12 AT&T Dataphone Digital Service Rate Evolution

Mileage Band	2.4 kbps		4.8 kbps		9.6 kbps		19.2 kbps		56 kbps	
	Fixed	Per Mile	Fixed	Per Mile	Fixed	Per Mile	Fixed	Per Mile	Fixed	Per Mile
A. In Effect Early 1990										
1–50	32.98	2.84	52.98	2.84	72.98	2.84	83.19	3.23	275.50	9.19
51–100	109.48	1.31	129.48	1.31	149.48	1.31	170.19	1.49	516.50	4.37
101–500	189.48	0.51	209.48	0.51	229.48	0.51	261.62	0.57	677.50	2.76
501+	284.48	0.32	304.48	0.32	324.48	0.32	366.26	0.36	1287.50	1.54
B. Tariff FCC No. 9, Effective April 20, 2001										
1–50	902.00	0.73	902.00	0.73	902.00	0.73	902.00	0.73	606.00	10.45
51–100	902.00	0.73	902.00	0.73	902.00	0.73	902.00	0.73	796.00	6.65
101–500	902.00	0.73	902.00	0.73	902.00	0.73	902.00	0.73	1111.00	3.50
501+	902.00	0.73	902.00	0.73	902.00	0.73	902.00	0.73	2062.00	1.60

Table 3.13　56 kbps DDS System Life Cost Computation

	Onetime Cost ($)	Monthly Recurring Cost ($)
Access lines	250	100
AT&T central office	250	100
Mileage		2686
DSUs (2 @ $200)	400	
	900	2886
System Life Cost Computation		
Onetime	900	
Recurring ($2886 × 36)	103,896	
System life cost	104,796	
Monthly cost (104,796/36)	2911	

3.4.5　Accunet T1.5

The 1.544 Mbps data transmission capability of Accunet T1.5 service provides a cost-effective method to integrate voice, data, and video transmission requirements between two locations. The use of Accunet T1.5, unlike DDS, will normally require the use of routers, T1 multiplexers, or channel banks to effectively use the available bandwidth. In addition, because the functions of a DSU are built into many routers and T1 multiplexers, users of Accunet T1.5 will have to use a separate CSU on a T1 line.

To illustrate the economic trade-offs between a 56 kbps DDS and a 1.544 Mbps T1 line, let us first examine AT&T's Accunet T1.5 service tariff. In doing so, we will first examine the tariff rate in effect in early 1990. Then, for comparison purposes, we will look at the AT&T Accunet T1.5 service rate in effect during 2003. It should be noted that due to deregulation, AT&T's currently available business and government offers that were formerly found in certain FCC tariffs can now be viewed online at AT&T Business Offers (http://serviceguide.att.com).

Table 3.14A lists the rate schedule for the circuit mileage between AT&T central offices in effect for Accunet T1.5 service during 1990. In comparison, Table 3.14B lists the AT&T business service rate for Accunet T1.5 service in late 2003.

In examining the evolution of Accunet T1.5 pricing, note that although the per-mile charge significantly decreased, the onetime fixed charge increased. If we consider the literal glut of fiber optics installed over the past few years, this can easily explain the decrease in per-mile charges because literally hundreds to thousands of T1.5 circuits can be transported

Table 3.14   Accunet T1.5 Service Rate Schedule and Monthly Recurring Charges

Circuit Miles	Fixed Charge ($)	Per-Mile Charge ($)
A. 1990 Tariff		
1–51	1800	10.00
51–100	1825	9.50
100+	2025	7.50
B. 2003 Business Service Rate		
	2300	3.85

over different types of fiber optics. Concerning the fixed charge, because this involves order processing and labor rates increased over the past decade it is natural to expect that cost element would increase.

3.4.5.1   DDS as Opposed to Accunet T1.5

We can obtain an indication of the cost-effectiveness of a T1 circuit by comparing the monthly cost of a 56 kbps DDS circuit to that of a T1 circuit. A T1 circuit can be used to support the transmission of twenty-four 56 or 64 kbps data sources, however, the monthly per-mile charge between a 56 kbps DDS and a 1.544 Mbps T1.5 circuit is almost equivalent for circuits up to 500 miles in length at current rates. At distances exceeding 500 miles, the monthly circuit cost of an Accunet T1.5 facility is less than twice the cost of 56 kbps digital service although the T1.5 service provides 24 times the capacity of the DDS service.

As with other system cost computations, you must consider the installation and monthly cost of access lines to each AT&T POP, as well as the cost of CSUs and routers or multiplexers or channel banks when using Accunet T1.5 facilities. Once you do so, you will probably determine the break-even point between separate 56 kbps DDS circuits and one Accunet T1.5 line to be between two and three DDS circuits. We say "probably" because of the number of variables that must be considered when developing the system cost of each facility, including operating company access line charges and DSU, CSU, and either router or multiplexer costs.

3.4.6   Accunet Spectrum of Digital Services

From a comparison of DDS and T1.5 monthly charges, it is obvious that end users that required the use of a few 56 kbps circuits were economically

Table 3.15 Accunet Spectrum of Digital Services Rate Schedule, Effective Late 2003

Data Rate (kbps)	Mileage Band	Fixed Monthly ($)	Per-Mile Monthly ($)
9.6	Any	1,021.00	0.81
56/64	Any	1,021.00	0.81
128	Any	1,838.00	1.47
192	Any	2,696.00	2.15
256	Any	3,513.00	2.80
320	Any	4,289.00	3.43
384	Any	5,024.00	4.00
448	Any	5,718.00	4.55
512	Any	6,372.00	5.06
576	Any	6,984.00	5.55
640	Any	7,557.00	6.02
704	Any	8,086.00	6.44
768	Any	8,577.00	6.82

forced to use a 1.544 Mbps facility, even though using that facility resulted in a large amount of unused bandwidth. To alleviate this situation, as well as to promote the use of digital transmission based upon groupings of one or more 64 kbps time slots in a T1 carrier facility, AT&T and other vendors introduced FT1 service. AT&T's FT1 service provides data transmission rates that are fractions of the 1.544 Mbps T1 transmission rate, ranging from a 56/64 kbps channel to a 768 kbps channel, as well as a low-speed 9.6 kbps facility.

3.4.6.1 Comparison to Other Digital Services

Table 3.15 contains AT&T's ASDS rate schedule that was in effect in late 2003. By comparing the mileage band rates between DDS, Accunet T1.5, and ASDS and their data transmission capability, we can examine the relative cost-performance ratio of each facility. Although all line charges are subject to change, we can obtain an indication of the cost-performance relationship between each of the 3 previously mentioned AT&T transmission offerings by computing the monthly cost of different length circuits based upon each 1000 bps of transmission provided by each type of transmission facility. Using the line costs contained in Table 3.12, Table 3.14, and Table 3.15 for 2003, we can compute the cost per 1000 bps of transmission for 5 types of DDS line facilities, 1.544 Mbps Accunet T1.5, and the 13 ASDS data rates for a 1000-mile circuit. Computing the monthly

**Table 3.16 Monthly Cost
per 1000 bps Transmission Capacity
for a 1000-Mile Interchange Circuit**

Digital Service (kbps)	Monthly Cost per 1000 bps ($)
DDS	
2.4	680.00
4.8	340.00
9.6	170.00
19.2	85.00
56	65.39
Accunet T1.5	3.98
ASDS	
9.6	190.16
56/64	32.96/28.60
128	25.84
192	25.23
256	24.66
320	24.12
384	23.50
448	22.91
512	22.33
576	21.76
640	21.21
704	20.63
768	20.05

cost of 1000-mile DDS circuits from Table 3.12 and dividing each cost by the data rate obtainable on the line in kbps results in the monthly cost per 1000 bps of transmission capacity. Thus, 2.4 kbps DDS would have a monthly cost of $680.00 per 1000 bps on a 1000-mile circuit. Similarly, a 1000-mile 56 kbps DDS circuit would have a monthly cost per 1000 bps of transmission capacity of $65.39.

Table 3.16 summarizes the monthly cost per 1000 bps of transmission for DDS, Accunet T1.5, and ASDS 1000-mile circuits. From Table 3.16, it is obvious that Accunet T1.5 service provides the best cost-performance ratio of all three of AT&T's digital transmission services that we just discussed. Next, all FT1 (ASDS) service offerings provide a cost-performance ratio more than approximately two times better than each DDS offering. Due to the disparity between ASDS and DDS, it is reasonable to

**Table 3.17 Monthly Line Cost Comparison —
56 kbps DDS versus 56/64 kbps ASDS**

Circuit Distance	56 kbps DDS ($)	56/64 kbps ASDS ($)	Ratio DDS/ASDS Monthly Cost ($)
1	616.45	1021.81	0.60
50	1128.50	1061.50	1.06
100	1461.00	1102.00	1.33
250	1986.00	1223.50	1.62
500	2861.00	1426.00	2.00
1000	3662.00	1831.00	2.00
2000	5262.00	2641.00	1.99

expect FT1 services to eventually replace a majority of AT&T's and other carriers' older digital transmission services.

From an examination of the costs associated with FT1 services, it is obvious that the spectrum of data transmission rates under ASDS can provide substantial savings for many users. These savings are most pronounced when a cost comparison to DDS occurs. In addition, significant cost savings are obtained when the rate for a full T1 circuit is compared to the cost of most 64 kbps groupings under ASDS.

To obtain a detailed appreciation for some of the abnormalities in digital transmission service tariffs, we will compare and contrast the cost of 56 kbps DDS to 56/64 kbps ASDS and 768 kbps ASDS to 1.544 Mbps Accunet T1.5 service. Table 3.17 contains the monthly cost comparison for 56 kbps DDS and 56/64 kbps ASDS circuits from 1 to 2000 miles in length.

Note that the fourth column in Table 3.17 indicates that the monthly cost ratio between DDS and 56/64 kbps ASDS rises to 2.00 for a circuit distance of 500 or more miles, illustrating the widening cost gap between the two services that provide an equivalent data transmission capacity. Due to this, it is reasonable to expect a growth in the availability of ASDS to be countered by a migration of users off DDS, unless the cost of usage for the latter is significantly reduced.

A second comparison of two digital service offerings produces unexpected results and involves 768 kbps ASDS and Accunet T1.5. Table 3.18 compares the monthly cost of 768 kbps ASDS to 1.544 Mbps Accunet T1.5 for circuit distances of 1, 50, 100, 250, 500, 1000, and 2000 miles.

In examining the entries in Table 3.18, note that for all circuit distances, the cost of a 768 kbps circuit actually exceeds the cost of a full T1 circuit operating at 1.544 Mbps. Thus, this pricing comparison represents one of the mysteries of life and shows why less may result in more cost.

**Table 3.18 Monthly Line Cost Comparison —
768 kbps ASDS versus Accunet T1.5**

Circuit Distance	768 kbps ASDS	1.544 Mbps T1.5	Ratio ASDS/T1.5 Monthly Cost
1	8583.82	2303.85	3.72
50	8918.00	2492.50	3.57
100	9259.00	2685.00	3.44
250	10,282.00	3262.50	3.15
500	11,987.00	4225.00	2.83
1000	15,397.00	6150.00	2.50
2000	22,217.00	10,000.00	2.22

3.4.6.2 Accunet T45

To conclude our discussion of leased digital facilities, we will briefly discuss the cost of a digital transmission facility that provides 28 times the capacity of a T1 line. AT&T's Accunet T45 service transmission facility provides support for a data rate of 44.736 Mbps.

During late 2003, the cost associated with an Accunet T45 interoffice channel on the AT&T Web site was based upon a fixed charge of $6100 per month and a per-mile cost of $21.00 per month. We can determine the number of T1 lines that will equal the cost of a T3 line on an interoffice basis. To do so, we would set the cost of x T1 lines to that of a T3 circuit for different distances. For example, consider a 100-mile interoffice series of T1 lines versus a T3 circuit. Then:

$$(2300 + 3.85 \times 100)x = 6100 + 21.00 \times 100$$

Solving for x, we obtain 3.05.

This means that if your organization needs the bandwidth associated with approximately three or more T1 circuits that will be routed between two common locations 100 miles apart, it might be economically advantageous to install a T3 circuit. We say "might be" because the preceding analysis did not include the cost of CSUs nor the local loop.

To illustrate the effect of interoffice mileage, assume the distance between locations was increased to 1000 miles. Now the break-even point between a number of T1 circuits and a T3 circuit becomes:

$$(2300 + 3.85 \times 1000)x = 6100 + 21.00 \times 1000$$

which results in a value of 4.4. Thus, as the interoffice distance increases, the number of T1 lines needed to equal the interoffice cost of a T3 line increases. Although it now takes 4.4 T1s to equal the interoffice cost of a T3 circuit, it is important to remember that the latter provides 28 times the transmission capacity of the former. Therefore, if your organization anticipates additional applications or an increase in usage and your computations indicate your current requirements are slightly below the break-even point, you may wish to consider ordering the higher capacity transmission facility to facilitate growth and to provide a more economical transmission method later on.

4

MULTIPLEXING, ROUTING, AND LINE SHARING EQUIPMENT

In this chapter, we will focus our attention upon the use of several types of multiplexers, routers, and line sharing devices that enable multiple data sources to be transmitted over a common WAN transmission line facility. The acquisition of each of the devices examined in this chapter is primarily justified by the economic savings and functionality their use promotes.

In the first section of this chapter, we will investigate the operation and use of several types of multiplexers. Once this is accomplished, we will turn our attention to a device originally developed to route data between LANs and which eventually acquired the name router in recognition of its basic design function. This will be followed by focusing attention upon a third category of line sharing equipment known by such terms as modem, line, and port sharing units. Although we will focus our attention upon the operation and use of multiplexing, routing, and line sharing equipment, we will defer detailed information concerning the methods that can be employed to determine the location or locations to install such equipment. In Chapter 6, we will examine the application of graph theory to several equipment location problems.

4.1 MULTIPLEXERS AND DATA CONCENTRATORS

With the establishment of distributed computing, the cost of providing the required communications facilities became a major concern of users. Numerous network structures were examined to determine the possibilities of using specialized equipment to reduce these costs. For many networks where geographically distributed users accessed a common computational facility, a central location could be found to serve as a hub to link those users to the computer. Even when terminal traffic was low and the cost of leased lines could not be justified on an individual basis, quite often the cumulative cost of providing communications to a group of users could be reduced if a mechanism was available to enable many terminals to share common communications facilities. This mechanism was originally provided by using multiplexers whose primary function is to provide the user with a reduction of communications costs. This device enables one high-speed line to be used to carry the formerly separate transmissions of a group of lower speed lines. Using multiplexers should be considered when a number of data terminals communicate from within a similar geographical area or when a number of leased lines run in parallel for any distance.

4.1.1 Evolution

From the historical perspective, multiplexing technology can trace its origination to the early development of telephone networks. Then, as today, multiplexing was the employment of appropriate technology to permit a communications circuit to carry more than one signal at a time.

In 1902, 26 years after the world's first successful telephone conversation, an attempt to overcome the existing ratio of one channel to one circuit occurred. Using specifically developed electrical network terminations, three channels were derived from two circuits by telephone companies.

The third channel was denoted as the phantom channel, hence the name "phantom" was applied to this early version of multiplexing. Although this technology permitted two pairs of wires to effectively carry the load of three, the requirement to keep the electric network finely balanced to prevent crosstalk limited its practicality.

4.1.2 Comparison with Other Devices

In the past, differences between multiplexers and concentrators were pronounced. Because multiplexers are prewired, fixed logic devices, they produced a composite output transmission by sharing frequency bands (frequency division multiplexing or FDM) or time slots (time division multiplexing or TDM) on a predetermined basis, with the result that the total transmitted output was equal to the sum of the individual data inputs. Multiplexers were also originally transparent to the communicator, so that data sent from a terminal through a multiplexer to a computer was received in the same format and code by the computer as its original form.

In comparison, concentrators were developed from minicomputers by the addition of specialized programming and originally performed numerous tasks that could not be accomplished through the use of a multiplexer. First, the intelligence provided by the software in concentrators permits a dynamic sharing technique used in traditional multiplexers. If a terminal device connected to a concentrator is not active, then the composite high-speed output of the concentrator will not automatically reserve a space for that terminal as will a traditional multiplexer.

This scheme, commonly known as dynamic bandwidth allocation, permits a larger number of terminal devices to share the use of a high-speed line through the use of a concentrator than when such terminals are connected to a multiplexer, because the traditional multiplexer allocates a time slot or frequency band for each terminal, regardless of whether the terminal is active. For this reason, statistics and queuing theory play an important role in the planning and use of concentrators. Next, owing to the stored program capacity of concentrators, these devices can be programmed to perform a number of additional functions. Such functions as the preprocessing of sign-on information and code conversion can be used to reduce the burden of effort required by the host computer system.

The advent of statistical and intelligent multiplexers, which are discussed later in this section, closed the gap between concentrators and multiplexers. Through the use of built-in microprocessors, these multiplexers could be programmed to perform numerous functions previously available only through the use of concentrators. In fact, many vendors marketed products labeled as data concentrators that were based upon the use of built-in microprocessors. Perhaps even more interesting is that several manuals described those concentrators as devices that perform statistical multiplexing.

The proliferation of LANs and the need to interconnect such local networks gave rise to a new communications device that was originally known as a gateway due to its initial function, which provided a mechanism for stations on one LAN to access those on another. Over the years, the near ubiquities of LAN at commercial organizations, government agencies, and academia, as well as the growth in the Internet, resulted in the evolution of the router's functionality. Although some routers continue to provide a simple gateway function, other devices now include such features as Voice-over–IP (VoIP), quality of service (QoS), and expedited forwarding techniques that enabled many end-user organizations to replace their multiplexers and data concentrators with routers. Although routers have over 90 percent of the market for transmitting multiple data sources over a common transmission line facility, many organizations continue to use high-speed multiplexers. In addition, in an evolutionary process many routers incorporate some of the functionality developed for multiplexers; therefore, we will discuss multiplexers prior to covering routers in this chapter.

Table 4.1   Candidates for Data Stream Multiplexing

Analog network private line modems
Analog switched network modems
Digital network data service units
Digital network channel service units
Data terminals
Data terminal controllers
Digitized voice
Minicomputers
Concentrators
Computer ports
Computer–computer links
Other multiplexers

4.1.3. Device Support

In general, any device that transmits or receives a serial data stream can be considered a candidate for multiplexing. Data streams produced by the devices listed in Table 4.1 are among those that can be multiplexed. The intermix of devices, as well as the number of any one device whose data stream is considered for multiplexing, is a function of the multiplexer's capacity and capabilities, the economics of the application, and cost of other devices that could be employed in that role, as well as the types and costs of high-speed lines being considered.

4.1.4   Multiplexing Techniques

Two basic techniques are used for multiplexing — FDM and TDM. Within the time division technique, two versions are available — fixed time slots, which are employed by traditional TDMs, and variable-use time slots, which are used by statistical and intelligent TDMs.

4.1.4.1   Frequency Division Multiplexing

In the FDM technique, the available bandwidth of the line is split into smaller segments called data bands or derived channels. Each data band in turn is separated from another data band by a guard band, which is used to prevent signal interference between channels, as shown in Figure 4.1. The guard band actually represents a group or band of unused frequencies. Because frequency drift is the main cause of signal interference and the size of the guard bands is structured to prevent data in one channel drifting into another channel, the guard band can be considered to represent a protection mechanism.

Figure 4.1   FDM Channel Separations
In FDM, the 3 kHz bandwidth of a voice-grade line is split into channels of data bands separated from each other by guard bands.

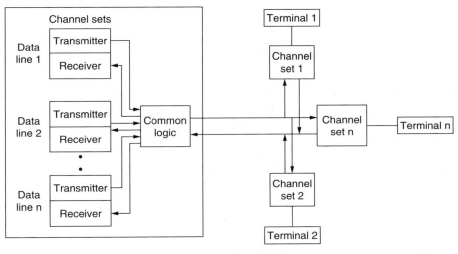

Figure 4.2   Frequency Division Multiplexing
Because the channel sets modulate the line at specified frequencies, no modems are required at remote locations.

Although many readers may consider FDM to represent an obsolete technology, to paraphrase Mark Twain, its demise is greatly exaggerated. In fact, two of the more popular modern implementations of FDM include DSL and cable modems. Both devices subdivide the frequency available on the medium into subbands for transmission and reception of data.

Physically, a FDM designed for use on voice-grade circuits contains a channel set for each data channel, as well as common logic, as shown in Figure 4.2. Each channel set contains a transmitter and receiver tuned to a specific frequency, with bits being indicated by the presence or absence of signals at each of the channel's assigned frequencies. In FDM, the width of each frequency band determines the transmission rate capacity

Table 4.2   FDM Channel Spacings for Low-Speed Transmissions

Speed (bps)	Spacing (Hz)
75	120
110	170
150	240
300	480
450	720
600	960
1200	1800

of the channel and the total bandwidth of the line is a limiting factor in determining the total number or mix of channels that can be serviced. Although a multipoint operation is illustrated in Figure 4.2, FDM equipment can also be used to multiplex data between two locations on a point-to-point circuit. Data rates up to 1200 bps can be multiplexed by a voice-grade FDM. Typical FDM channel spacings that are required at different data rates for multiplexing data onto a voice-grade analog line are listed in Table 4.2.

As indicated in Table 4.2, the channel spacing on an FDM system is proportional to its data rate. In the 1960s and early 1970s when terminal transmission rates were relatively low, FDM could provide for the multiplexing of a fairly reasonable number of data sources over the 3000 Hz bandwidth of an analog leased line. As terminal transmission speeds increased, the number of channels that could be supported by FDM significantly decreased, making this multiplexing technique essentially obsolete for modern data transmission over WANs. Although DSL and cable modems use FDM, they do so within a local area, such as from a subscriber to a telephone company central office in the case of a DSL. Because this reduces network design to a series of point-to-point connections for DSL and a tree structure for cable modems, the technology is not used in a WAN environment. Thus, the remainder of this section will focus its attention upon several types of TDM.

4.1.4.2 Time Division Multiplexing

In the TDM technique, the aggregate capacity of the line is the frame of reference, because the multiplexer provides a method of transmitting data from many terminal devices over a common circuit by interleaving data from each device by time. The TDM divides the aggregate transmission on the line for use by the slower speed devices connected to the multi-

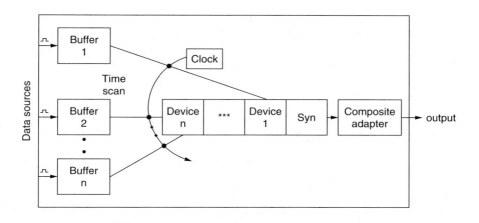

Figure 4.3   Time Division Multiplexing
In TDM, data is first entered into each channel adapter buffer area at a transfer rate equal to the device to which the adapter is connected. Next, data from the various buffer areas are transferred to the multiplexer's central logic at the higher rate of the device for packing into a message frame for transmission.

plexer. Each device is given a time slot for its exclusive use so that at any one point in time the signal from one terminal is on the line.

The fundamental operating characteristics of a TDM are shown in Figure 4.3. Here, each low- to medium-speed data source is connected to the multiplexer through an input/output (I/O) channel adapter. The I/O adapter provides the buffering and control functions necessary to interface the transmission and reception of data to the multiplexer. Within each adapter, a buffer or memory area exists that is used to compensate for the speed differential between the data sources and the multiplexer's internal operating speed. Data is shifted to the I/O adapter at different rates, depending upon the speed of the connected devices; but when data is shifted from the I/O adapter to the central logic of the multiplexer, or from central logic to the composite adapter, it is at the much higher fixed rate of the TDM. On output from the multiplexer to each connected device the reverse is true, because data is first transferred at a fixed rate from central logic to each adapter and then from the adapter to the connected device at the data rate acceptable to the device. Depending upon the type of TDM system, the buffer area in each adapter will accommodate either bits or characters.

The central logic of the TDM contains controlling, monitoring, and timing circuitry that facilitates the passage of individual data sources to and from the high-speed transmission medium. The central logic will generate a synchronizing pattern that is used by a scanner circuit to interrogate each of the channel adapter buffer areas in a predetermined sequence, blocking the bits of characters from each buffer into a continuous, synchronous data stream, which is then passed to a composite adapter. The composite adapter contains a buffer and functions similar to the I/O channel adapters. However, it now compensates for the difference in speed between the high-speed transmission medium and the internal speed of the multiplexer.

4.1.5 The Multiplexing Interval

When operating, the multiplexer transmits and receives a continuous data stream known as a message train, regardless of the activity of the data sources connected to the device. The message train is formed from a continuous series of message frames, which represents the packing of a series of input data streams. Each message frame contains one or more synchronization characters followed by a number of basic multiplexing intervals whose number is dependent upon the model and manufacturer of the device. The basic multiplexing interval can be viewed as the first level of time subdivision, which is established by determining the number of equal sections per second required by a particular application. Then, the multiplexing interval is the time duration of one section of the message frame.

When TDMs were first introduced, the section rate was established at 30 sections per second, which then produced a basic multiplexing interval of 0.033 second or 33 ms. Setting the interval to 33 ms made the multiplexer directly compatible to a 300-baud asynchronous channel, which transmits data at up to 30 characters per second (cps). With this interval, the multiplexer was also compatible with 150-baud (15-cps) and 110-baud (10-cps) data channels, because the basic multiplexing interval was a multiple of those asynchronous data rates. Later TDMs had a section rate of 120 sections per second, which then made the multiplexer capable of servicing a range of asynchronous data streams up to 1200 bps. As developments in Universal Asynchronous Receiver/Transmitter (UART) technology enabled asynchronous terminal devices to achieve greater data rates, multiplexers were redesigned to support higher section rates, with section rates of 1920 and 3840 introduced to support data rates up to 19,200 bps and 38,400 bps, and some multiplexers were designed to support even higher asynchronous data rates.

4.1.6 TDM Techniques

The two TDM techniques available are bit interleaving and character interleaving. Bit interleaving is generally used in systems that service synchronous terminal devices, whereas character interleaving is generally used to service asynchronous terminal devices. When interleaving is

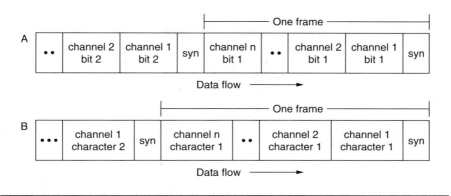

Figure 4.4 Time Division Interleaving — Bit-by-Bit and Character-by-Character
When interleaving is accomplished bit by bit (A), the first bit from each channel is packed into a frame for transmission. When interleaving is conducted on a character-by-character basis (B), one or more complete characters are grouped with a synchronization character into a frame for transmission.

accomplished on a bit-by-bit basis, the multiplexer takes one bit from each channel adapter and then combines them as a word or frame for transmission. As shown in Figure 4.4A, this technique produces a frame containing one data element from each channel adapter. When interleaving is accomplished on a character-by-character basis, the multiplexer assembles a full character from each data source into one frame and then transmits the frame, as shown in Figure 4.4B.

For the character-by-character method, the buffer area required is considerably larger. This means that additional time is required to assemble a frame when multiplexing occurs on a character-by-character basis than when a TDM bit interleaving method is employed. However, because the character-by-character interleaved method preserves all bits of a character in sequence, the TDM equipment can be used to strip the character of any recognition information that may be sent as part of that character. Examples of this would be the servicing of such terminal devices as a Teletype™ Model 33 or a personal computer (PC) transmitting asynchronous data, where a transmitted character contains 10 or 11 bits, which include 1 start bit, 7 data bits, 1 parity bit, and 1 or 2 stop bits. When the bit interleaved method is used, all 10 or 11 bits would be transmitted to preserve character integrity, whereas in a character interleaved system, the start and stop bits can be stripped from the character, with only the 7 data bits and on some systems the parity bit warranting transmission. This means that a character-by-character interleaving method can be significantly more efficient than a bit-by-bit interleaving method, although the former introduces more delay.

To service terminal devices with character codes containing different numbers of bits per character, two techniques are commonly employed in character interleaving. In the first technique, the time slot for each character is of constant size, designed to accommodate the maximum bit width or highest level code. Making all slots large enough to carry American Standard Code for Information Interchange (ASCII) characters makes the multiplexer an inefficient carrier of a lower level code, such as five-level Baudot. The second technique used is to proportion the slot size to the width of each character according to its bit size. This technique maximizes the efficiency of the multiplexer, although the complexity of the logic and the cost of the multiplexer increases. Owing to the reduction in the cost of semiconductors, most character interleaved multiplexers were designed to operate on the proportional assignment method.

Although bit interleaving is simple to perform, it is also less efficient when used to service asynchronous terminals. On the positive side, bit interleaved multiplexers offer the advantage of faster resynchronization and shorter transmission delay, because character interleaved multiplexers must wait to assemble the bits into characters; whereas, a bit interleaved multiplexer can transmit each bit as soon as it is received from the terminal device. Multiplexers, which interleave by character, use a number of different techniques to build the character, with the techniques varying between manufacturers and by models produced by manufacturers.

A commonly used technique is the placement of a buffer area for each channel adapter, which permits the character to be assembled within the channel adapter and then scanned and packed into a data stream. Another technique that can be used is the placement of programmed read-only memory (ROM) within the multiplexer so that it can be used to assemble characters for all the input channels. The second technique makes a multiplexer resemble a data concentrator, because the inclusion of pro-grammed ROM permits many additional functions to be performed in addition to the assembly and disassembly of characters. Such multiplexers with programmed memory are referred to as intelligent multiplexers and formed the foundation for the development of many router features. Both communications devices are discussed later in this chapter.

4.1.7 TDM Constraints

Prior to investigating some of the more common network applications for which TDMs are used, we should examine the major constraints that limit the number of channels supported by a TDM and the data transmission rates obtainable on those channels. To do so, we will examine a typical TDM configuration from both a hardware and a network schematic con-figuration perspective. The top of Figure 4.5 illustrates an 8-channel TDM

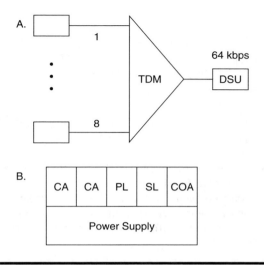

Figure 4.5   Network and Hardware TDM Schematics
CA — channel adapter card, PL — primary logic card, SL — secondary logic card, COA — composite adapter card

connected to a DSU operating at 64 kbps as viewed from a network perspective. In the lower portion of that illustration, the 8-channel TDM is illustrated from a hardware perspective, assuming that the equipment is mounted in an industry standard 19-inch rack.

In examining the network schematic, let us focus our attention upon the data rates the TDM can support. Although eight inputs are illustrated, no data rates were actually assigned to individual channels, because prior to doing so you must consider several device constraints.

The first constraint you must consider is the data rate supported by each channel adapter card. Most multiplexers have channel adapters that support 2, 4, or 8 line terminations that are referenced as multiplexer ports or channels. Because the operational characteristics of the channel adapter govern the operational characteristics of each of the line terminations mounted in the adapter, in effect, all ports on the adapter have the same data rate and protocol support characteristics. Thus, if one port supports 300, 1200, 2400, and 4800 bps asynchronous transmission, all other ports on the adapter can be expected to have a similar support capability.

The actual data rate each port can be set to depends upon several factors in addition to the data rates supported by the channel adapter housing the port. These factors include the aggregate data input supported by the TDM, the data rate assigned to other ports, the data rate assigned to the composite adapter, and the overhead and efficiency of the TDM.

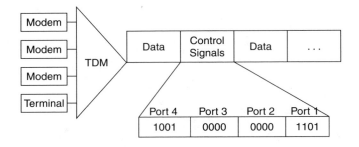

Figure 4.6 Multiplexer Overhead
Once every *x* data sample cycles, a TDM will sample the state of a number of RS-232 control leads on each port, encoding them as a binary 0, if not active, or binary 1, if active. Passing control signals is the major contribution to the overhead of TDMs.

Concerning the next to last factor, overhead is primarily a result of the sampling of control signals on the input interfaces. For example, if a multiplexer supported RS-232 interfaces, the multiplexer would have to sample the state of the RS-232 interface once every predefined number of data sample cycles. In addition, on the composite, output gaps in transmission between groups of bits or characters will arise from the need to send synchronization characters from one multiplexer to another.

When a multiplexer performs the sampling of the RS-232 interface, it typically examines multiples of two leads on each port and encodes the state of the leads as a binary value — zero if the lead is low and one if the lead is high. In this manner, the multiplexer can pass the state of many control leads to a distant location. As an example, this enables a ring indicator signal passed by an auto-answer modem connected to a multiplexer or a data terminal ready signal caused by a directly connected terminal being powered on to be passed to a distant computer system as illustrated in Figure 4.6.

To illustrate the effect of control signal passing upon the overhead of a TDM, let us assume every 25th cycle the device samples control signals. If the composite adapter is connected to a 64 kbps DSU, then 1/25 of 64,000 or 2560 bps is used for the transmission of control signals, precluding the transmission of data. Due to this, the maximum data rate that all input ports could accommodate without the TDM losing data would be 64,000 − 2560 or 61,440 bps, assuming there were no other overhead factors or constraints to consider. Because the transmission rates supported by each port are usually a multiple of 300 bps, the aggregate data input of all ports would probably be limited to int(64,000/300) × 300 or 63,900 without considering control signal and any other overhead.

The protocol overhead associated with multiplexers can be expected to lower the actual data transmission of TDMs by another 10 to 15 percent. Continuing with our previous example, this could lower the effective data transmission servicing capacity of the TDM from 61,440 bps to 57,300 bps or less.

Although many vendors provide detailed information concerning data rate constraints and protocol overhead associated with their TDMs, some vendors only include this information in manuals that are received after you purchase the equipment. Thus, it is imperative that you check the previously mentioned constraints prior to purchasing a TDM to ensure its overhead will not adversely affect your network application.

Counteracting the overhead of TDMs are several operational efficiencies that, upon occasion, negate the overhead of the device. Two of the more common operational efficiencies are the use of a proportional width time slot and the stripping of nonessential bits from asynchronous characters, such as start, stop, and parity bits. If vendor literature does not provide detailed information concerning the aggregate data rate their equipment can support based upon a specific composite adapter operating rate, you can estimate that rate with a fair degree of accuracy if you know the overhead and efficiency factors of the TDM. As an example of this, again assume the TDM uses 1 cycle every 25 cycles for the passing of control information and its composite adapter operates at 64 kbps. Then, the overhead is 2560 bps. If we assume that the TDM is to service six 9600 bps terminal devices, without considering any possible efficiencies provided by the multiplexer, our first impression would be that the device would be limited to supporting five such devices. This is because 6 × 9600 is 57,600 bps, however, as previously computed, less than 64,000 bps of data can be effectively transmitted as 2560 bps is reserved for control signaling and 6400 bps or more for the protocol overhead. Now let us investigate the effect of the TDM stripping start and stop bits prior to transmission while the distant multiplexer reconstructs each character by the addition of those bits prior to transmitting them to their destination.

A 9600 bps data source using 10 bits per character presents 960 characters per second to a multiplexer port. If the multiplexer strips each start and stop bit, only 8 bits per character are actually transmitted over the multiplexed line. Thus, the effective data rate per port is 960 cps × 8 bits per character or 7680 bps. If all six multiplexer ports transmit similar ASCII characters whose start and stop bits can be stripped, the effective data input rate becomes 7680 x 6 or 46,080 bps. This rate is low enough to be serviced by the composite adapter's effective operating rate, which was computed by considering the passing of control characters and the overhead associated with the protocol used by the multiplexers.

The hardware schematic illustrated in the lower portion of Figure 4.5 will be used to discuss some of the physical constraints associated with TDMs, as well as similar equipment. In that illustration, note that there are five cards inserted into the upper bay or rack-mount nest. Most rack-mount equipment has housings that include a bus built into the backplane of the nest, as well as slots for the insertion of adapter cards. Similar to a personal computer, the insertion of adapter cards adds to the functionality and capability of the TDM.

If each channel adapter is capable of supporting four ports, the hardware schematic illustrated in Figure 4.5 indicates that this particular TDM can support a maximum of eight data sources. To expand the capability of the TDM, you can consider several options depending upon the method by which the device is manufactured. First, you might be able to remove the secondary logic card and manually use it to replace a failed primary logic card, if the need arises. Then, you could install a third channel adapter card to increase the number of data sources supported to 12. Another option would be to install a third bay, adding five more slots into which different adapter cards can be inserted. When selecting this option, you must consider the power requirements of the additional adapter cards that could require an additional power supply to operate. Thus, you must consider the number of data sources each adapter card supports, the number of adapter cards that can be installed in a bay, as well as the number of bays and the power requirements to support adapter cards installed in each bay.

4.1.8 TDM Applications

Although the focus of this section is upon TDMs, many of the architectural configurations shown are also applicable to routers. This will become more apparent in the next section when we turn our attention to that device.

The most commonly used TDM configuration is the point-to-point system, which is shown in Figure 4.7. This type of system, which is also called a two-point multiplex system, links a mixture of terminal devices to a centrally located multiplexer. As shown, the terminals can be connected to the multiplexer in a variety of ways. Terminals can be connected by a leased line running from the terminal's location to the multiplexer, by a direct connection if the user's terminals are within the same building as the multiplexer and a cable can be laid to connect the two, or terminals can use the switched network to call the multiplexer over the dial network. For the last method, because the connection is not permanent, several terminals can share access to one or more multiplexer channels on a contention basis.

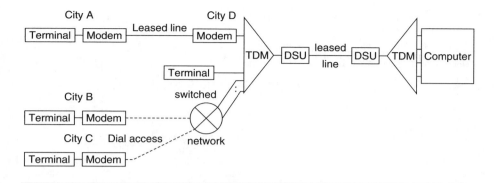

Figure 4.7   TDM Point to Point
A point-to-point or two-point multiplexing system links a variety of data users at one or more remoted locations to a central computer facility.

As shown in Figure 4.7, the terminals in City B and City C use the dial network to contend for multiplexer channels that are interfaced to automatic answering modems connected to the dial network. Whichever terminal accesses a particular channel maintains use of it and thus excludes other terminals from access to that particular connection to the system. As an example, one might have a network that contains 50 terminals within a geographical area, wherein between 10 to 12 are active at any time. One method to deal with this environment would be to install a 12-number rotary or hunt group interfaced to a 12-channel multiplexer. If all of the terminals were located within one city, the only telephone charges that the user would incur in addition to those of the leased line between multiplexers would be local call charges each time a terminal user dialed the local multiplexer number.

4.1.8.1   Series Multipoint Multiplexing

A number of multiplexing systems can be developed by linking the output of one multiplexer into a second multiplexer. Commonly called series multipoint multiplexing, this technique is most effective when terminal devices are distributed at two or more locations and the user desires to alleviate the necessity of obtaining two long-distance leased lines from the closer location to the computer. As shown in Figure 4.8, four low-speed terminals are multiplexed at City A onto one high-speed channel, which is transmitted to City B where this line is in turn multiplexed along with the data from a number of other terminals at City B. Although the user requires a leased line between City A and City B, only one line is now required to be installed for the remainder of the distance from City B to the computer at City C. If City A is located 50 miles from City B and City

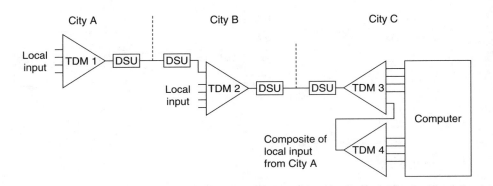

Figure 4.8 Series Multipoint Multiplexing
Series multipoint multiplexing is accomplished by connecting the output of one multiplexer as input to a second device.

B is 2000 miles from City C, 2000 miles of duplicate leased lines are avoided by using this multiplexing technique.

Multipoint multiplexing requires an additional pair of channel cards to be installed at Multiplexer 2 and Multiplexer 3 and higher speed modems or DSUs to be interfaced to those multiplexers to handle the higher aggregate throughput when the traffic of Multiplexer 1 is routed through Multiplexer 2. But, in most cases, the cost savings associated with reducing duplicated leased lines will more than offset the cost of the extra equipment. Because this is a series arrangement, a failure of either TDM2 or TDM3, either high-speed modem or DSU, or a failure of the line between these two multiplexers will terminate service to all terminals connected to the system.

4.1.8.2 Hub-Bypass Multiplexing

A variation of series multipoint multiplexing is hub-bypass multiplexing. To be effectively used, hub-bypass multiplexing can occur when a number of remote locations are required to transmit to two or more locations. To satisfy this requirement, the remote terminal traffic is multiplexed to a central location that is the hub and the terminals that must communicate with the second location are cabled into another multiplexer that transmits this traffic, bypassing the hub. Figure 4.9 illustrates one application where hub-bypassing might be used. In this example, eight terminals at City 3 require a communications link with one of two computers, six terminals always communicate with the computer at City 2, while two terminals use the facilities of the computer at City 1. The data from all eight terminals are multiplexed over a common line to City 2 where the two channels

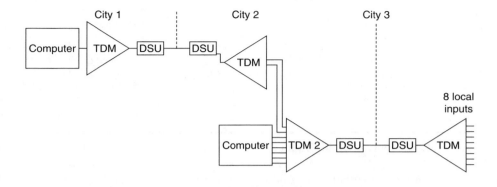

Figure 4.9   Hub-Bypass Multiplexing
When a number of terminals are required to communicate with more than one location, hub-bypass multiplexing should be considered.

Figure 4.10   TDM System Used as a Front End
When a TDM is used as a front-end processor, the computer must be programmed to perform demultiplexing.

that correspond to the terminals that must access the computer at City 1 are cabled to a new multiplexer, which then re-multiplexes the data from those terminals to City 1. When many terminal locations have dual location destinations, hub-bypassing can become very economical. However, because the data flows in series, an equipment failure will terminate access to one or more computational facilities, depending upon the location of the break in service.

Although hub-bypass multiplexing can be effectively used to connect collocated terminals to different destinations, if more than two destinations exist a more efficient switching arrangement can be obtained by employing a router, as we will note in the next section.

4.1.8.3   Front-End Substitution

Although not commonly used, a TDM may be installed as an inexpensive front end for a computer, as shown in Figure 4.10.

When used as a front end, only one computer port is required to service a group of terminal devices that are connected to the TDM. The TDM can be connected at the computer center or it can be located at a

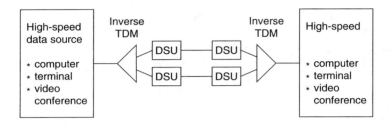

Figure 4.11   Inverse Multiplexing
An inverse multiplexer splits a serial data stream into two or more individual data streams for transmission at lower data rates.

remote site and connected over a leased line and a pair of modems or DSUs. Because demultiplexing is conducted by the computer's software, only one multiplexer is necessary.

However, owing to the wide variations in multiplexing techniques of each manufacturer, no standard software has been written for demultiplexing. Unless multiple locations can use this technique, the software development costs can easily exceed the hardware savings associated with this technique. In addition, the software overhead associated with the computer performing the demultiplexing may degrade its performance to an appreciable degree and must be considered.

4.1.8.4   Inverse Multiplexing

A multiplexing system that is coming into widespread usage is the inverse multiplexing system, which is also referred to when using the dial-up digital network as a bandwidth-on-demand multiplexer. As shown in Figure 4.11, inverse multiplexing permits a high-speed data stream to be split into two or more slower data streams for transmission over lower cost lines and modems or DSUs.

Because of the tariff structure associated with wideband facilities, using inverse multiplexers can result in significant savings in certain situations. As an example, their use could permit 112,000 bps transmission over two voice-grade lines at a fraction of the cost that would be incurred when using wideband facilities.

For an example of the economics associated with the potential use of inverse multiplexers, refer to the monthly line cost comparison between 19.2 and 56 kbps DDS leased lines presented in Chapter 3. From Table 3.12B, you will note that a 2000-mile 19.2 kbps digital leased line has a cost of $2362 per month and a 56 kbps DDS circuit has a cost of $5262 per month. Thus, the use of two 19.2 lines could provide an aggregate transmission capacity of 38.4 kbps if each line operates at 19.2 kbps for a total line cost of $4724 per month. Then, if the cost of a pair of inverse multiplexers and four DSUs is less than the cost of two digital service units used on a 56 kbps DDS circuit plus the difference between the cost of one 56 kbps digital line and two 19.2 kbps circuits, inverse multiplexing will be more economical. In addition, two lower speed 19.2 kbps lines can serve as a backup to one another enabling transmission at 19.2 kbps to occur if one circuit should become inoperative. In comparison, the failure of the single 56 kbps DDS circuit would cause the transmission system to become inoperative.

4.1.9 Multiplexing Economics

The primary motive for the use of multiplexers in a network is to reduce the cost of communications. Until the development of the modern router, most organizations operated a mainframe facility at a central location with remote users at branch offices requiring access to the central computer. In this section, we will turn our attention to the economic aspects associated with the use of multiplexers, deferring coverage of routers to Section 4.2.

In analyzing the potential of multiplexers, one should first survey terminal users to determine the projected monthly connect time of each terminal. Then, the most economical method of data transmission from each individual terminal device to the computer facility they wish to access can be computed. To do this, direct dial costs should be compared with the cost of a leased line from each terminal device to the computer site.

Once the most economical method of transmission for each individual terminal to the computer is determined, this cost should be considered the cost to reduce. The telephone mileage costs from each terminal city location to each other terminal city location should be determined to compute and compare the cost of using various techniques, such as line dropping and multiplexing data by combining several low- to medium-speed terminals' data streams into one high-speed line for transmission to the central site.

In evaluating multiplexing costs, the cost of telephone lines from each terminal location to the multiplexer center must be computed and added to the cost of the multiplexer equipment. Then, the cost of the high-speed line from the multiplexer center to the computer site must be added to produce the total multiplexing cost. If this cost exceeds the cumulative most economical method of transmission for individual terminals to the central site, then multiplexing is not cost-justified. This process should be reiterated by considering each city as a possible multiplexer center to optimize all possible network configurations. In repeating this process, terminals located in certain cities will not justify any calculations to prove or disprove their economic feasibility as multiplexer centers, because of their isolation from other cities in a network.

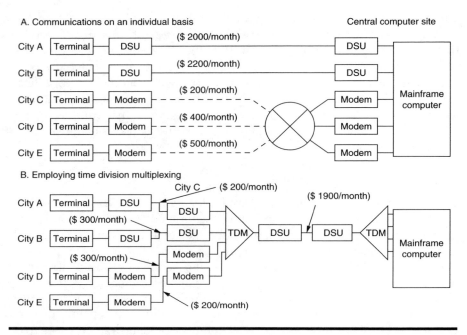

Figure 4.12 Multiplexing Economics
On an individual basis, the cost of five terminals accessing a computer system (A) can be much more expensive than if TDM is installed (B).

An example of the economics involved in multiplexing is illustrated in Figure 4.12. In this example, assume the volume of terminal traffic from the terminal devices located in City A and City B would result in a dial-up charge of $3000 per month per location if access to the computer in City G was over the switched network. The installation of leased lines from those cities to the computer at City G would cost $2000 and $2200 per month, respectively. Thus, it would be more economical to install a leased line from City A and City B directly to City G if we did not consider the potential use of multiplexers. Furthermore, let us assume that the terminals at City C, City D, and City E only periodically communicate with the computer and their dial-up costs of $200, $400, and $500 per month, respectively, are much less than the cost of leased lines between those cities and the computer. Then, without multiplexing, the network's most economical communications cost and type of access to the computer at City G would be as shown in Table 4.3.

Let us further assume that City C is centrally located with respect to the other cities, so we could use it as a homing point or multiplexer center. In this manner, a multiplexer could be installed in City C and the terminal traffic from the other cities could be routed to that city, as shown

Table 4.3   Most Economical Communications Cost

Location	Cost per Month	Type of Access
City A	$2000	Leased line
City B	2200	Leased line
City C	200	Switched network
City D	400	Switched network
City E	500	Switched network
Total cost	$5300	

in Figure 4.12B. Employing multiplexers would reduce the network communications cost to $2900 per month, which produces a potential savings of $2400 per month, which should now be reduced by the multiplexer costs to determine net savings. If each multiplexer costs $500 per month, then the network using multiplexers will save the user $1400 each month. Exactly how much can be saved, if any, through the use of multiplexers depends not only on the types, quantities, and distributions of terminals to be serviced but also on the leased line tariff structure and the type of multiplexer employed.

4.1.10   Statistical and Intelligent Multiplexers

In a traditional TDM, data streams are combined from a number of devices into a single path so that each device has a time slot assigned for its exclusive use. Although such TDMs are inexpensive and reliable and can be effectively employed to reduce communications costs, they make inefficient use of the high-speed transmission medium. This inefficiency is due to the fact that a time slot is reserved for each connected device, whether or not the device is inactive. When the device is inactive, the TDM pads the slot with nulls or pad characters and cannot use the slot for other purposes.

These pad characters are inserted into the message frame because demultiplexing occurs by the position of characters in the frame. Thus, if these pads are eliminated, a scheme must then be employed to indicate the origination port or channel of each character. Otherwise, there would be no way to correctly reconstruct the data and route it to its correct computer port during the demultiplexing process.

A statistical multiplexer is, in many respects, similar to a concentrator and can be considered as the predecessor to the modern router. This is because each of these communications devices combines signals from a number of connected devices in such a manner that there is a certain probability that a device will have access to the use of a time slot for transmission. Whereas a concentrator may require user programming and always requires special software in the host computer to demultiplex its high-speed data stream, statistical multiplexers are built around a microprocessor that is programmed by the vendor and no host software is required for demultiplexing, because another statistical multiplexer at the computer site performs that function. When we discuss routers in this chapter, we will note how they resemble statistical multiplexers as they only form packets for transmission when data is received from an input device and use a microprocessor programmed by the vendor to perform a series of predefined operations on data.

By dynamically allocating time slots as required, statistical multiplexers permit more efficient use of the high-speed transmission medium. This permits the multiplexer to service more terminal devices without an increase in the high-speed link as would a traditional multiplexer. The technique of allocating time slots on a demand basis is known as statistical multiplexing and means that data is transmitted by the multiplexer only from the terminal devices that are actually active.

Depending upon the type of TDM, either synchronization characters or control frames are inserted into the stream of message frames. Synchronization characters are employed by conventional TDMs, but control frames are used by TDMs that employ a HDLC protocol or a version of that protocol between multiplexers to control the transmission of message frames.

The construction technique used to build the message frame also defines the type of TDM. Conventional TDMs employ a fixed frame approach as illustrated in Figure 4.13. Here, each frame consists of one character or bit for each input port or channel scanned at a particular period of time. As illustrated, even when a particular terminal is inactive, the slot assigned to that device is included in the message frame transmitted, because the presence of a pad or null character in the time slot is required to correctly demultiplex the data. The lower portion of Figure 4.13 illustrates the demultiplexing process accomplished by time slot position. Because a typical data source may be idle 90 percent of the time, this technique contains obvious inefficiencies.

4.1.10.1 Statistical Frame Construction

A statistical multiplexer (STDM) employs a variable frame building technique that takes advantage of terminal idle times to enable more terminals to share access to a common circuit. The use of variable frame technology permits previously wasted time slots to be eliminated, because control information is transmitted with each frame to indicate which terminals are active and have data contained in the message frame.

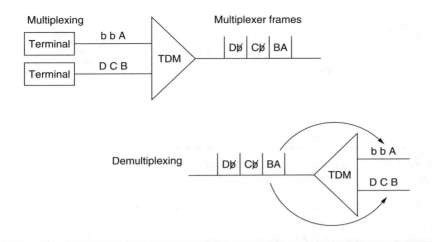

Figure 4.13 Multiplexing and Demultiplexing by TDMs
Legend: b = absence of activity during multiplexer scan; b̷ = null character inserted into multiplexer message frame.

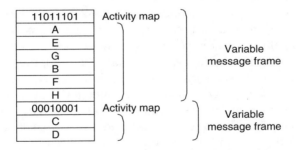

Figure 4.14 Activity Mapping to Produce Variable Frames
Using an activity map, where each bit position indicates the presence or absence of data for a particular data source, permits variable message frames to be generated.

One of many techniques that can be used to denote the presence or absence of data traffic is the activity map illustrated in Figure 4.14. When an activity map is employed, the map itself is transmitted before the actual data. Each bit position in the map is used to indicate the presence or absence of data from a particular multiplexer time slot scan. The two activity maps and data characters illustrated in Figure 4.14 represent a total of 10 characters that would be transmitted in place of the 16 characters if no activity mapping occurred.

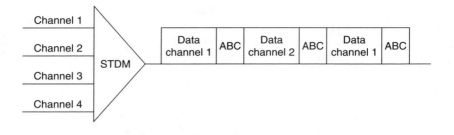

Figure 4.15 Address and Byte Count Frame Composition
Legend: ABC = address and byte count.

Another statistical multiplexing technique involves buffering data from each data source and then transmitting the data with an address and byte count. The demultiplexer uses the address to route the data to the correct port and the byte count indicates the quantity of data to be routed to that port.

Figure 4.15 illustrates the message frame of a four-channel statistical multiplexer employing the address and byte count frame composition method during a certain time interval. Note that because Channel 3 and Channel 4 had no data traffic during the two particular time intervals, there was no address and byte count nor data from those channels transmitted on the common circuit. Also note that the data from each channel is of variable length. Typically, statistical multiplexers employing an address and byte count frame composition method wait until either 32 characters or a carriage return is encountered prior to forming the address and byte count and forwarding the buffered data. The reason 32 characters was selected as the decision criterion is that it represents the average line length of an interactive transmission session.

Users should note that a few potential technical drawbacks of statistical multiplexers exist. These problems include the delays associated with data blocking and queuing when a large number of connected terminal devices become active or when a few devices transmit large bursts of data. For either situation, the aggregate throughput of the multiplexer's input active data exceeds the capacity of the common high-speed line, causing data to be placed into buffer storage.

Another reason for delays is when a circuit error causes one or more retransmissions of message frame data to occur. Because the connected terminal devices may continue to send data during the multiplexer-to-multiplexer retransmission cycle, this can also fill up the multiplexer's buffer areas and cause time delays.

If the buffer area should overflow, data would be lost. This would create an unacceptable situation, except in the wonderful world of routers where data loss is a common occurrence. To prevent buffer overflow, all statistical multiplexers employ some type of technique to transmit a traffic control signal, formerly referred to as flow control, to attached terminal devices when their buffers are filled to a certain level. Such control signals inhibit additional transmission through the multiplexer until the buffer has been emptied to another predefined level. Once this level has been reached, a second control signal is issued that permits transmission to the multiplexers to resume.

4.1.10.2 Buffer Control

The three major buffer control techniques employed by statistical multiplexers include inband signaling, outband signaling, and clock reduction. Inband signaling involves transmitting XOFF and XON characters to inhibit and enable the transmission of data from terminals and computer ports that recognize these flow control characters. Because many terminals and computer ports do not recognize these control characters, a second common flow control method involves raising and lowering the Clear to Send (CTS) control signal on the RS-232 or CCITT V.24 interface. Because this method of buffer control is outside the data path where data is transmitted on Pin 2, it is known as outband signaling.

Both inband and outband signaling are used to control the data flow of asynchronous devices. Because synchronous devices transmit data formed into blocks or frames, one would most likely break a block or frame by using either inband or outband signaling. This would cause a portion of a block or frame to be received, which would result in a negative acknowledgment when the receiver performs its cyclic redundancy computation. Similarly, when the remainder of the block or frame is allowed to resume its flow to the receiver, a second negative acknowledgment would result.

To alleviate these potential causes of throughput decrease, multiplexer vendors typically reduce the clocking speed furnished to synchronous devices. Thus, a synchronous terminal device operating at 64,000 bps might first be reduced to 32,000 bps by the multiplexer halving the clock. Then, if the buffer in the multiplexer continues to fill, the clock might be further reduced to 24,000, 16,000, or even a lower data rate.

4.1.10.3 Service Ratio

The measurement used to denote the capability of an STDM is called its service ratio, which compares its overall level of performance to a conventional TDM. Because synchronous transmission by definition denotes blocks of data with characters placed in sequence in each block, there

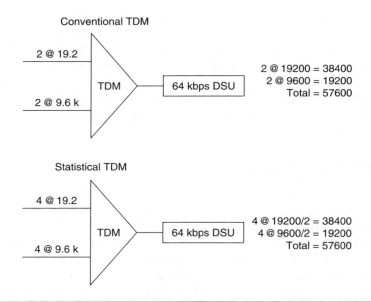

Figure 4.16 Comparing Statistical and Conventional TDMs
An STDM typically has an efficiency of two to four times a conventional TDM. Using an efficiency level twice the conventional TDM results in the ability to support an input composite operating data rate double that of a conventional TDM.

are no gaps in this mode of transmission. In comparison, a terminal operator transmitting data asynchronously may pause between characters to think prior to pressing each key on the terminal. Thus, the service ratio of STDMs for asynchronous data is higher than the service ratio for synchronous data. Typically, STDM asynchronous service ratios range between 2:1 and 3.5:1, and synchronous service ratios range between 1.25:1 and 2:1, with the service ratio dependent upon the efficiency of the STDM, as well as its built-in features, including the stripping of start and stop bits from asynchronous data sources. In Figure 4.16, the operational efficiency of both a statistical and conventional TDM are compared. Here, we have assumed that the STDM has an efficiency of twice that of the TDM.

Assuming two 19,200 bps and two 9600 bps data sources are to be multiplexed, the conventional TDM illustrated in the top part of Figure 4.16 would be required to operate at a data rate of at least 57,600 bps, thus requiring a 64,000 bps DSU if transmission was over a digital facility. For the STDM shown in the lower portion of this illustration, assuming a two-fold increase in efficiency over the conventional TDM, the composite

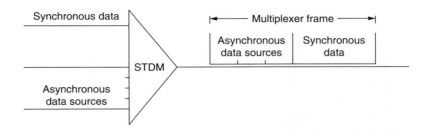

Figure 4.17   Using a Bandpass Channel to Multiple Synchronous Data

input data rate the device can support is doubled, in this example to 104,400 bps.

4.1.10.4   Data Source Support

Some STDMs only support asynchronous and synchronous data sources. When a statistical multiplexer supports synchronous data sources, it is extremely important to determine the method used by the STDM vendor to implement this support.

Some STDM vendors employ a bandpass channel to support synchronous data sources. When this occurs, not only is the synchronous data not multiplexed statistically, but the data rate of the synchronous input limits the overall capability of the device to support asynchronous transmission. Figure 4.17 illustrates the effect of multiplexing synchronous data via the use of a bandpass channel. When a bandpass channel is employed, a fixed portion of each message frame is reserved for the exclusive multiplexing of synchronous data, with the portion of the frame reserved proportional to the data rate of the synchronous input to the STDM. This means that only the remainder of the message frame is then available for the multiplexing of all other data sources.

As an example of the limitations of bandpass multiplexing, consider an STDM that is connected to a 64,000 bps DSU and supports a synchronous terminal device operating at 56,000 bps. If bandpass multiplexing is employed, only 8000 bps is then available in the multiplexer for the multiplexing of other data sources. In comparison, assume another STDM statistically multiplexes synchronous data. If this STDM has a service ratio of 1.5:1, then a 56,000 bps synchronous input to the STDM would, on the average, take up 37,334 bps of the 64,000 bps operating line. Because the synchronous data is statistically multiplexed, when that data source is not active other data sources serviced by the STDM will flow through the system more efficiently. In comparison, the bandpass channel always

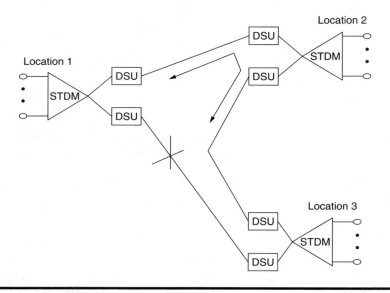

Figure 4.18   Data Switching
Switching permits load balancing and alternate routing if a high-speed line should become inoperative.

requires a predefined portion of the high-speed line to be reserved for synchronous data, regardless of the activity of the data source.

4.1.10.5   Switching and Port Contention

Two features normally available with more sophisticated STDMs are switching and port contention. Switching capability is also referred to as alternate routing and requires the multiplexer to support multiple high-speed lines whose connection to the multiplexer is known as a node. Thus, switching capability normally refers to the ability of the multiplexer to support multiple nodes. Figure 4.18 illustrates how alternate routing can be used to compensate for a circuit outage. In the example shown, if the line connecting Location 1 and Location 3 should become inoperative, an alternate route through Location 2 could be established if the STDMs support data switching.

Although many organizations no longer use multiplexers, many of their features served as development goals by router manufacturers. In fact, some of the more common multiplexer features we will cover later in this section are incorporated into routers, resulting in many persons considering the TDM as the forerunner of the modern router.

As second feature incorporated into sophisticated STDMs that warrants discussion is port contention. Port contention is normally incorporated into large capacity multimodal STDMs that are designed for installation at a central computer facility. This type of STDM may demultiplex data from hundreds of data channels, however, because many data channels are usually inactive at a given point in time, it is a waste of resources to provide a port at the central site for each data channel on the remote multiplexers. Thus, port contention results in the STDM at the central site containing a lesser number of ports than the number of channels of the distant multiplexers connected to that device. Then, the STDM at the central site contends the data sources entered through remote multiplexer channels to the available ports on a demand basis. If no ports are available, the STDM may issue a NO PORTS AVAILABLE message and disconnect the user or put the user into a queue until a port becomes available.

4.1.10.6 Intelligent Time Division Multiplexes

One advancement in STDM technology resulted in the introduction of data compression into STDMs. Such devices intelligently examine data for certain characteristics and are known as intelligent time division multiplexers (ITDM). These devices take advantage of the fact that different characters occur with different frequencies and use this quality to reduce the average number of bits per character by assigning short codes to frequently occurring characters and long codes to seldom-encountered characters.

The primary advantage of ITDMs lies in their ability to make the most efficient use of a high-speed data circuit in comparison to the other classes of TDMs. Through compression, synchronous data traffic that normally contains minimal idle times during active transmission periods can be boosted in efficiency. ITDMs typically permit an efficiency four times that of conventional TDMs for asynchronous data traffic and twice that of conventional TDMs for synchronous terminals. Thus, in analyzing the traffic handling capacity of ITDMs, you can usually work with higher service ratios than those available with STDMs.

4.1.10.7 STDM/ITDM Statistics

Although the use of STDMs and ITDMs can be considered on a purely economic basis to determine if the cost of such devices is offset by the reduction in line and modem or DSU costs, the statistics that are computed and made available to the user of such devices should also be considered. Although many times intangible, these statistics may warrant consideration

Table 4.4   Intelligent Multiplexer Statistics

Multiplexer loading = percent of time device not idle
Buffer utilization = percent of buffer storage in use
Number of frames transmitted
Number of bits of idle code transmitted
Number of negative acknowledgments received
Traffic density = nonidle bits/total bits
Error density = NAKs received/frames transmitted
Compression efficiency = total bits received/total bits compressed
Statistical loading = number of actual characters received/maximum number
 which could be received
Character error rate = characters with bad parity/total characters received

even though an economic benefit may at first be hard to visualize. Some of the statistics normally available on STDMs and ITDMs are listed in Table 4.4. Through a careful monitoring of these statistics, network expansion can be preplanned to cause a minimum amount of potential busy conditions to users. In addition, frequent error conditions can be noted prior to user complaints and remedial action taken earlier than normal when conventional devices are used.

4.1.10.8   Using System Reports

One of the most common problems associated with the use of an STDM or ITDM service ratio is that vendors providing this figure tend to exaggerate, using a best-case scenario to compute the ratio. As an example, vendors might consider all asynchronous users to be performing interactive data entry functions and ignore the possibility that some users are performing full screen emulated access to a mainframe via a protocol converter. Similarly, vendors will probably ignore file transfer activities, because they would also increase data transmission and lower the computation of the multiplexer's service ratio.

One obvious question to ask is "what is the effect of connecting too many data sources to an STDM or ITDM based upon an inflated service ratio?" To understand the effect resulting from this situation, let us examine the cause and effect of buffer control on a terminal operator's display.

Figure 4.19 illustrates an STDM for which a service ratio of 4:1 for asynchronous transmission is assumed. In this illustration, assume all 20 multiplexer ports are connected to mainframe computer ports. As more and more terminal users become actively connected to computer ports through the STDM system, begin to pull files, scroll through screens, or perform other data transmission intensive operations, the data flow into

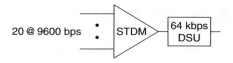

Figure 4.19   Overconfigured STDM
As all ports become fully active transmitting data into the STDM, the data flow into the device approaches 192 kbps; however, the data flow from the STDM to the line is 64 kbps. This causes the STDM's buffer to fill rapidly.

the multiplexer rapidly increases and could conceivably approach 192 kbps. Because the multiplexer transfers data onto the line at 64 kbps, the difference between the input and output data flow will rapidly fill the STDM's buffer. To prevent buffer overflow, the multiplexer will implement buffer or flow control when the buffer fills to a predetermined level.

When the multiplexer implements flow control, it signals the computer ports to disable or stop data transmission, providing the multiplexer with the ability to dump data out of its buffers. At another predefined buffer level, the STDM will issue another set of buffer or flow control signals to enable transmission to resume.

If a large number of file transfer operations are in progress, the multiplexer will rapidly enable and disable transmission again and again. To typical terminal operators connected to the computer through the STDM system, the multiplexer's actions will result in random pauses in the display of data on their screen. In addition to obviously affecting response time, the random nature of the pauses will adversely affect the productivity of terminal operators and result in numerous complaints. These complaints are usually directed at the host computer's capacity as the symptoms closely resemble the effect of a saturated computer system. Only after hardware capacity planners review the use of the mainframe, do most organizations begin to examine the possibility that their multiplexers are overloaded.

One of the best mechanisms available to determine potential multiplexer overloading conditions and initiate corrective action is through the use of system reports generated by most multiplexers. System reports generated by STDMs and ITDMs can include the parameters listed in Table 4.4 as well as other data elements. In this section, we will examine the use of several system report parameters to understand how this information can be used to accurately size data sources to the capacity of STDMs and ITDMs.

Table 4.5 illustrates an example of a multiplexer Line Status Report. This report is generated at predefined intervals and indicates the inbound and outbound character rates per second averaged over the prior interval

Table 4.5   Typical Line Status Report

HH:MM:SS LINE 64,000 BPS

Outbound	10 MIN AVG 7574
Inbound	10 MIN AVG 2289

The typical multiplexer line status report indicates the average inbound and outbound traffic with respect to the line operating rate between two multiplexers in the system.

of time. In addition, some multiplexers also provide the peak character input rate over a smaller period of time.

As indicated in Table 4.5, the high-speed line is operating at 64,000 bps, which is equivalent to 8000 cps. Thus, over the prior 10-minute period, data being transmitted through the local multiplexer to the remote multiplexer (outbound) represents a line utilization of 7574/8000 × 100 or 94.67 percent. Similarly, data being transmitted from the remote site through the remote multiplexer (inbound) represents a line occupancy rate of 2289/8000 × 100 or 28.61 percent.

As a rule of thumb, any line occupancy in excess of 80 to 85 percent sustained over a period of time will result in an excessive amount of flow control. This is because the line occupancy calculations based upon inbound and outbound data compared to the line capacity rate do not include the effect of the overhead associated with the protocol used to transfer data between multiplexers. Typically, the protocol overhead may utilize 10 to 15 percent or more of the line's total traffic handling capacity. Thus, without examining other report parameters, a line occupancy report can be used to determine whether or not your multiplexer system has or is approaching an overload condition.

If line occupancy is excessive, you can consider one of several options to improve multiplexer performance. First, you can replace existing modems or DSUs on the high-speed connections with higher speed devices to increase the transmission rate on the line, in effect, reducing the line occupancy. If it is not feasible to increase the transmission rate between multiplexers, you can consider removing one or more data sources that may be excessively using the capacity of the multiplexer, as they may warrant an individual communications circuit. In some cases, the identification of more active data sources can be facilitated by the reports some multiplexers provide that indicate inbound and outbound traffic on an individual port basis, as well as on a composite basis.

Another report element available from some multiplexers includes the number of flow controls issued by each channel during the predefined

Table 4.6   Typical Multiplexer Buffer Report

HH:MM:SS LINE 64,000 BPS

LOCAL TRANSMIT BUFFER	10 MIN AVG 1842 CHAR
REMOTE TRANSMIT BUFFER	10 MIN AVG 436 CHAR

The buffer report can be used to determine throughput delays through multiplexers. This information can be used to determine what effect multiplexers have on composite response time delays seen from the terminal operator's perspective.

interval of time. If this report element is available, you can directly determine which ports are being adversely affected and use this information to take corrective action.

A third report element provided by some multiplexers indicates the occupancy of its transmit buffers over a predefined period of time. Table 4.6 indicates an example of this report element that can be a separate multiplexer report or incorporated into a multiplexer's system report.

In effect, the buffer report indicates the number of characters contained in the multiplexer awaiting transmission onto the high-speed line. By comparing the average contents of the local and remote transmit buffers with the high-speed line transmission rate, you can determine inbound and outbound delays through the multiplexer system. This information is extremely useful in determining the contributing factors that may result in an unacceptable response time for interactive terminal users.

To illustrate the use of multiplexer buffer occupancy information, consider the local transmit buffer 10-minute average of 1842 characters shown in Table 4.6. This is equivalent to 14,736 bits, which, when transmitted at 64,000 bps, requires approximately 0.43 seconds until the first character entering the transmit buffer is placed onto the line. This means that outbound data traffic through the multiplexer is adding almost a half second to the total response time end users see between pressing an Enter or Return key on their terminal and having the first character of the computer's response displayed. Concerning the remote transmit buffer, note that its occupancy results in approximately five-hundredths of a second delay. Normally, if the local multiplexer is considered to be located at the computer site and the remote multiplexer is located where terminal users are located, you will find a similar disproportion between local and remote buffer occupancy. This disproportion is due to terminal queries being relatively short in comparison to computer responses generated by those queries. Thus, in most cases, you will probably concentrate your efforts toward determining methods that can reduce the occupancy of the transmit buffer of the local multiplexer.

Table 4.7   Statistical Multiplexer Selection Features

Feature	Parameters to Consider
Auto baud detect	Data rates detected
Flyback delay	Settings available
Echoplex	Selectable by channel or device
Protocols supported	2780/3780, 3270, HDLC/SDLC, other
Data type supported	Asynchronous, synchronous
Service ratios	Asynchronous, synchronous
Flow control	XON-XOFF, CTS, clocking
Multinodal capability	Number of nodes
Switching	Automatic or manual
Port contention	Disconnect or queued when all ports in use
Data compression	Stripping bits or employs compression algorithm

4.1.10.9   Features to Consider

In Table 4.7, you will find a list of the primary selection features to consider when evaluating STDMs. Although many of these features were previously discussed, a few features — auto baud detect, flyback display, and echoplex — were purposely omitted from consideration until now. These features primarily govern the type of terminal devices that can be efficiently supported by the STDM.

Auto baud detect represents the ability of a multiplexer to measure the pulse width of a data source. Because the data rate is proportional to the pulse width, this feature enables the multiplexer to recognize and adjust to different speed terminals accessing the device over the PSTN.

On electromechanical printers, a delay time is required between sending a carriage return to the terminal and then sending the first character of the next line to be printed. This delay time enables the print head of the terminal to be repositioned prior to the first character of the next line being printed. Many STDMs can be set to generate a series of fill characters after detecting a carriage return, enabling the print head of an electromechanical terminal to return to its proper position prior to receiving a character to be printed. This feature is called flyback delay and can be enabled or disabled by channel on many multiplexers.

Because some networks contain full-duplex computer systems that echo each character back to the originating terminal, the delay from twice traversing through STDMs may result in the terminal operator feeling that his or her terminal is nonresponsive. When echoplexing is supported by an STDM, the multiplexer connected to the terminal immediately echoes each character to the terminal, while the multiplexer connected to the computer discards characters echoed by the computer. This enables data flow through the multiplexer system to be more responsive to the terminal operator. Because error detection and correction is built into all STDMs, a character echo from the computer is not necessary to provide visual transmission validation and is safely eliminated by echoplexing. The other options listed in Table 4.7 should be self-explanatory and the user should check vendor literature for specific options available for use on different devices.

4.2 ROUTERS

In this section, we will turn our attention to a communications device that many persons consider as the successor to various types of TDMs. That device is the router, which evolved over the years from a simple product used to interconnect two geographically separated LANs into a sophisticated communications device that can transport digitized voice, data, and video, recognizing priorities within the headers of packets that govern how fast or slow the router switches packets from one port to another.

4.2.1 Functionality

All routers are designed to perform two common functions — switching packets and determining an optimal routing path. The basic switching of packets occurs when a router receives an inbound packet on one port. The router's software examines the header of the packet to determine its destination address. The router then examines its routing table, which associates learned addresses with destination networks and uses information in the table to switch the packet so that it flows outbound on a specific port.

Figure 4.20 illustrates a simple example of the use of a router routing table. In this example, a two port router is used to connect an Ethernet network (connected to the router's port E0) to the Internet via the router's serial port (port S0). Routers are initially configured with the network address associated with each port of the device. Thereafter, they learn other network addresses by transmitting various types of packets between devices. Such packets include information concerning which network addresses are associated with each port, enabling routers to build a picture of the structure of the interconnected network. Once that picture is made, routers can perform a second function referred to as path determination. Under path determination, routers can select the best path between source and destination based upon such metrics as line quality, delay, and other factors.

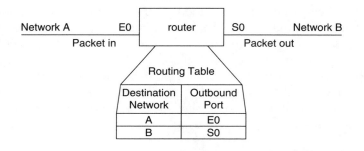

Figure 4.20 Two Basic Router Functions — Packet Switching and Path Determination

4.2.2 Ports and Connectors

Routers are modular devices, typically having two or more built-in ports, that expand their capability through the addition of adapter cards inserted into a chassis, similar to the manner by which adapter cards are added to a PC's system unit to increase the functionality of a computer. Router adapter cards typically support the addition of 1, 2, 4, or 8 ports. Each port, to include those built into the device, has a specific type of interface connector. For example, a serial port could have an RS-232 interface for supporting relatively low data rates or a V.35 interface to support data rates up to approximately 6 Mbps, with a high-speed serial interface (HSSI) used to support data transfer onto T3 circuits operating at approximately 45 Mbps. Similarly, LAN ports will have an interface connector designed to support a specific type of media, such as an RJ-45 connector for twisted-pair wiring.

4.2.3 Address Resolution

Routers switch packets while operating at Layer 3 in the Open Systems Interconnection (OSI) Reference Model. In comparison, when connected to a LAN, routers transmit and receive data in the form of frames that operate at Layer 2 in the OSI Reference Model.

When data flows on a LAN it uses 48-bit Media Access Control (MAC) addresses to identify the originator (source) and recipient (destination) of frames. This represents a slight problem when a router receives a packet from another network destined to a station on a connected LAN. The problem is the fact that the router knows the Layer 3 destination address, which is contained in the header of the packet, but needs to form a frame to deliver the packet using a Layer 2 destination address. Thus, the router needs to determine the Layer 2 address for which it knows the Layer 3 address.

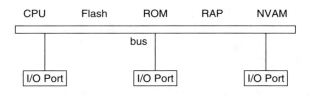

Figure 4.21   Router Hardware Components

The solution to the above addressing problem is in the form of the use of the Address Resolution Protocol (ARP). When a router needs to determine the Layer 2 address of a station for which it knows the Layer 3 address it transmits an ARP broadcast. The ARP broadcast flows to each station on the LAN other than the originating device. The broadcast contains the Layer 3 address for which it requires the corresponding Layer 2 address. The station with the defined Layer 3 address responds to the broadcast, returning its Layer 2 MAC address to the router, allowing that device to form a frame with the correct destination address to deliver the packet within a Layer 2 frame.

Because an ARP broadcast precludes the transfer of other data on the LAN, a mechanism is needed to reduce such broadcasts. That mechanism is in the form of an ARP cache, where the router stores learned MAC addresses that correspond to specific Layer 3 addresses. Thus, the router will actually check its ARP cache prior to transmitting an ARP broadcast.

4.2.4   Hardware and Software

Previously, we noted that the primary functions of a router are to switch packets and determine a path for data transfer. In addition to those two functions, routers have the capability to perform additional functions based upon their hardware and software. Thus, in this section, we will turn our attention to both topics.

Figure 4.21 illustrates the major hardware components of a typical router. Similar to an STDM, the central processing unit (CPU) of the router represents a microprocessor. ROM contains power-on diagnostics similar to the Power-On Self-Test (POST) included on a PC's ROM. In addition, the router's ROM can contain a bootstrap program that loads a copy of the operating system into memory upon power-on.

The router's random-access memory (RAM) is used to store routing tables, the ARP cache, and the router's configuration information. In addition, RAM is used as a buffer and queue as packets are switched from one port to another. In comparison, nonvolatile RAM (NVRAM) provides "permanent" storage for a router's configuration because information is retained when the router is powered off.

A third type of memory used in routers is flash memory. Flash memory can be considered to represent erasable, reprogrammable ROM that is used to store microcode that updates ROM, as well as one or more operating system images. When a router is powered on, it will execute any self-test in ROM and then use the bootstrap loader to copy an operating system image in RAM. That image can be stored in flash memory or located on a network device, with the exact location specified by the setting of a router's configuration register. Once the operating system is loaded, the router then loads a previously created configuration file from NVRAM or enables the administrator to configure the router. Thus, let us turn our attention to router software components.

4.2.4.1 Software Modules

All routers have two basic software modules — an operating system and a configuration. The operating system governs the manner by which the router supports different protocols, updates routing tables, manages buffers, switches packets between ports, and executes user commands. In comparison, the router configuration defines how the operating system supports the specific router platform. For example, the configuration can define the operating rates and protocol used on specific ports, the network address of those ports, and source or destination addresses that can transfer information into or out of those ports. The last activity is a router access list and represents one of many types of configuration data used by routers.

Because the operating system rarely varies, but the configuration can frequently change, configuration information is usually stored in several locations. First, its typically stored in NVRAM and loaded into the router's main memory each time the router is powered on. Because the configuration can be quite complex for large routers, resulting in hundreds to thousands of lines of data, several persons may be assigned to configure a router. By placing the configuration on a network accessible file, it then becomes possible for several persons to access and modify the configuration. In fact, in an operational environment, an organization may have several router configuration files. One file might be used for daily operations and a second file might store a test configuration that is applied during the weekend or in the evening to test one or more configuration changes prior to the new configuration moving into a routine use status.

4.2.5 Data Flow and Packet Switching

In concluding our introductory discussion of routers, we will turn our attention to the flow of data within a router and techniques developed to expedite the data flow. In doing so, we will use Figure 4.22 as a reference, because it depicts the general flow of data within a router.

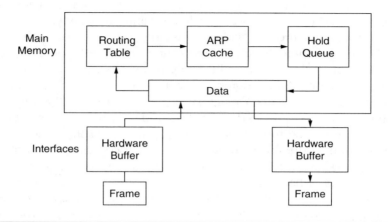

Figure 4.22   Data Flow within a Router

Data arrives at a router's interface in the form of either a LAN frame at a LAN port or a data link packet at a WAN port. The router's interface logic verifies the frame or packet format and performs a data integrity check by computing a cyclic reliability check (CRC) and comparing it to the CRC appended to the frame or packet. Assuming the two CRCs are equal, the router accepts the frame or packet. If the frame or packet format is not correct or the CRCs do not match, the router simply drops or discards the data and generates an error message depending upon the protocol used at the interface.

Assuming the frame or packet is accepted, it is buffered in the interface and an interrupt is sent to the microprocessor requesting service. The microprocessor executes a service routine, moving the frame or packet into main memory, with the location and status of the data tracked by an area in memory referred to as a hold queue. If the router is operating on a packet with a Layer 3 header, it compares its destination address to entries in the routing table to determine the outbound port for the packet. The router may use the ARP cache to determine the type of encapsulation required for the outbound frame and then switch it to the outbound port. That port adds heading and trailer information and outputs the packet as a serial data stream. If the router is operating on a Layer 2 frame, it removes the Layer 2 header to examine the packet being transported on the LAN and then performs the previously described series of operations.

The key metric associated with routers is its switching speed, commonly expressed in packets per second (pps). Over the years, a number of techniques were developed to enhance switching performance. Most techniques involve the use of route caches that contain information about destination addresses associated with ports. Entries in route caches are added as packets are forwarded through the switch, which means that initially packets take longer to be switched, because initially the tables are empty.

Router manufacturers provide various methods to enhance the switching efficiency of their devices. Some routers provide a route cache in a buffer area on their interface, alleviating the necessity of packets to flow into main memory. Other routers place a routing cache in main memory. Although the use of a main memory cache may appear to require more switching time than an interface cache, updating router interfaces using a route cache can be more time consuming than a main memory update.

In addition to the use of route caches, protocols have been modified or new protocols were developed to expedite switching. For example, the IPv4 protocol had a portion of its header modified to use eight bits as a type of service (ToS) byte to expedite the flow of marked packets. Obviously, a router needs to recognize such markings to expedite the flow of marked packets.

In concluding this chapter, we will briefly examine the use of equipment primarily associated with older and for the most part obsolete controller based terminals. However, because one person's obsolescence is another person's bargain and such equipment is still used in many locations throughout the world, let us turn our attention to different types of sharing devices.

4.3 MODEM AND LINE SHARING UNITS

Cost-conscious company executives are always happy to hear of ways to save money on the job. One of the things a data communications manager can do to make his or her presence felt is to produce a realistic plan for reducing expenses. It may be evident that a single communications link is less costly than two or more. What is sometimes less obvious is the most economical and effective way to make use of even a single link.

Multiplexing or routing are usually the first techniques that come to mind, but there are many situations where far less expensive, albeit somewhat slower, equipment is quite adequate. Here, terminals are polled one by one through a sharing device that acts under the instructions of the host computer. Typically, the applications where this method would be most useful and practical would be those where messages are short and where most traffic between host computer and terminal moves in one direction during any one period of time. The line-sharing technique (as distinct from multiplexing) may work in some interactive situations, but only if the overall response time can be kept within tolerable limits. The technique is not usually useful for remote batching or remote job entry (RJE), unless messages can be carefully scheduled so as not to get in each other's way, because of the long runtime for any one job. Although

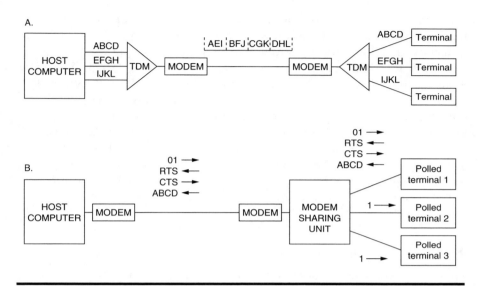

Figure 4.23 Multiplexing versus Line Sharing
A multiplexer needs a TDM system (A), which requires one computer port for each terminal and a multiplexer at each end. A sharing system (B) needs only one computer port. Because it requires terminals to be polled, a sharing system can be cost-effective for interactive operation, but may not be so for long messages that are likely to move in remote job entry or remote batch types of applications.

line sharing is inexpensive, it has some limits to its usefulness, particularly in situations where a multiplexer or router can be used to produce additional economic leverage.

4.3.1 Operation

A TDM operates continuously to sample in turn each channel feeding it, either bit by bit or character by character. This produces an aggregate transmission at a speed equal to the sum of the speeds of all its terminals.

A conventional TDM operation is illustrated in Figure 4.23A. For example, a multiplexer operating character by character assembles its first frame by taking the letter A from the first terminal, the letter E from the second, and the letter I from the third terminal. During the next cycle, the multiplexer takes the second character of each message (B, F, and J, respectively) to make up its second frame. The sampling continues in this way until traffic on the line is reversed to allow transmission from the computer to the terminals. The demultiplexing side of the TDM (operating on the receiving side of the network) assembles incoming messages and distributes them to their proper terminals or computer ports.

A line-sharing network connects to the host computer by a local link, through which the host polls the terminals one by one. The central site transmits the address of the terminal to be polled throughout the network by way of the sharing unit. This is illustrated in Figure 4.23B. The terminal assigned this address (01 in the diagram) responds by transmitting an RTS signal to the computer, which returns a CTS to prompt the terminal to begin transmitting its message (ABCD in diagram). When the message is completed, the terminal drops its RTS signal and the computer polls the next terminal.

Throughout this sequence, the sharing device continuously routes the signals to and from the polled terminal and handles supporting tasks, such as making sure the carrier signal is on the line when the terminal is polled and inhibiting transmission from all terminals not connected to the computer.

4.3.2 Device Differences

There are two types of devices that can be used to share a polled line:

1. Modem-sharing units
2. Line-sharing units

They function in much the same way to perform much the same task — the only significant difference being that the line-sharing unit has an internal timing source, but a modem-sharing unit gets its timing signals from the modem it is servicing.

A line-sharing unit is mainly used at the central site to connect a cluster of terminals to a single computer port, as shown in Figure 4.24. However, it does play a part in remote operation, when a data stream from a remote terminal cluster forms one of the inputs to a line-sharing unit at the central site to make it possible to run with a less expensive single-port computer.

In a modem-sharing unit, one set of inputs is connected to multiple terminals or processors, as shown in Figure 4.24. These lines are routed through the modem-sharing unit to a single modem. Besides needing only one remote modem, a modem-sharing network needs only a single two-wire (for half-duplex) or four-wire (for full-duplex) communications link. A single link between terminals and host computer allows all of them to connect with a single port on the host, a situation that results in still greater savings.

If multiplexing were used in this type of application, the outlay would likely be greater, because of the cost of the hardware and the need for a dedicated host computer port for each remote device. A single modem-sharing unit, at the remote site, is all that is needed for a sharing system, but multiplexers come in pairs, one for each end of the link.

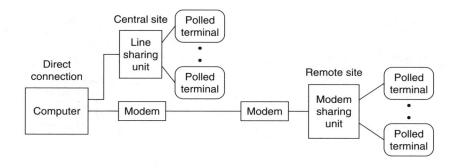

Figure 4.24   Line-Sharing and Modem-Sharing Use Compared
Line-sharing units tie central site terminals to the computer, but modem-sharing units handle all the remote terminals. A line-sharing unit requires internal timing, whereas a modem-sharing unit gets its timing from the modem to which it is connected. In either case, access to the host is made through a single communications link — either a two-wire or four-wire — and a single port at the central site computer.

The polling process makes sharing units less efficient than multiplexers. Throughput is cut back because of the time needed to poll each terminal and the line turnaround time on half-duplex links. Another problem is that terminals must wait their turn. If one terminal sends a long message, others may have to wait an excessive amount of time, which may tie up operators if unbuffered terminals are used; but terminals with buffers to hold messages waiting for transmission will ease this situation.

4.3.3   Sharing Unit Constraints

Sharing units are generally transparent within a communications network. There are, however, four factors that should be taken into account when making use of these devices:

1. Distance separating the data terminals and the sharing unit (for example, generally set at no more than 50 ft under RS-232-C interface specifications)
2. Number of terminals that can be connected to the unit
3. Types of modems with which the unit can be interfaced
4. Whether the terminals can accept external timing from a modem through a sharing unit

Then, in addition, the normal constraints of the polling process, such as delays arising from turnaround and response and the size of the transmitted blocks, must be considered in designing a sharing unit based network facility.

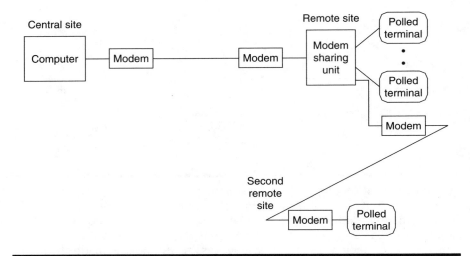

Figure 4.25 Extending the Connection
Line- or modem-sharing units form a single link between a host computer and terminals. This system contains a modem-sharing unit with inputs from the terminals at its own site, as well as from remote terminals. A line-sharing unit at the central site can handle either remote site devices or local devices more than 50 ft away from the host computer, which is the maximum cable length advisable under the RS-232-C/CCITT V.24 standards.

The 50-ft limit on the distance between terminal and sharing unit (RS-232/CCITT V.24 standard) can cause problems if terminals cannot be clustered closely. A way to avoid this constraint is to obtain a sharing unit with a data communications equipment (DCE) option. This option permits a remote terminal to be connected to the sharing unit through a pair of modems, as illustrated in Figure 4.25. This in turn allows the users the economic advantage of a through connection out to the farthest point. Because the advantage of modem-sharing units over a multipoint line is the reduction in the total number of modems when terminals are clustered, only one or at most a few DCE options should be used with a modem-sharing unit, as it could defeat the economics of clustering the terminals to use a common modem.

It is advisable to check carefully what types of modem can be supported by modem-sharing units, because some modems permit a great deal more flexibility of network design than others. For instance, if the sharing unit can work with a multipoint modem, the extra modem ports can service remote batch terminals or dedicated terminals that frequently handle long messages. An example of this flexibility of design is shown in Figure 4.26. Some terminals that cannot accept external timing can be

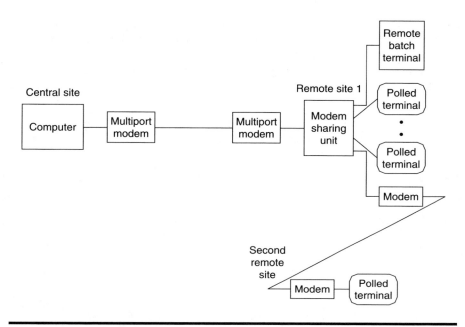

Figure 4.26   Multiple Applications Share the Line
Through the use of a modem-sharing unit with a data communications equipment interface, a terminal distant from the cluster (Location 2) can share the same line segment (computer to Location 1) that is used to transmit data to those terminals at Location 1. With a second application that requires a remote batch terminal at Location 1, additional line economies can be derived by installing multiport modems so both the polled terminals and the RBT can continue to share the use of one leased line from the computer to Location 1.

fitted with special circuitry through which the timing originates at the terminal itself instead of at the modem.

4.3.4   Summary

Modem- and line-sharing units were originally developed during the 1970s and 1980s to share the use of what were then expensive leased lines connecting regional and branch offices to corporate data centers. At that time, mainframe computer ports were expensive and the ability to both economize on leased line charges and reduce equipment costs could normally be expected to provide significant economic savings. The growth in local area networking reduced the previously popular mainframe centric networking structure and the need for sharing devices. In addition, IBM, which popularized the use of sharing units with its 3270 Information

Display System's series of control units several years ago, discontinued their manufacture. Today, you can still encounter both IBM and third-party sharing units in many data centers and their use continues to provide economic savings to organizations located around the globe.

5

LOCATING DATA CONCENTRATION EQUIPMENT

One of the most commonly encountered problems facing network designers is the selection of locations to install data concentration equipment. Although network designers must consider many factors, including the availability of space for equipment and personnel to provide maintenance vendors with access to such equipment, the major factor used to consider a location for data concentration equipment is economics. In general, when you have a number of locations transmitting and receiving data within a geographical area, you will attempt to select a location for the installation of data concentration equipment that represents a minimum cost of transmission from all locations within the geographical area to the selected location. Because graph theory can be used to provide a firm methodology for selecting a data concentration location that represents a minimum cost location with respect to all other locations, we will first examine this subject area. Using the relationship between graph theory and network topology will provide the information required to use graph theory to manually perform or automate the equipment location process, which is the major emphasis of the second section of this chapter.

5.1 GRAPH THEORY AND NETWORK DESIGN

One of the more interesting aspects of data communications networks is their topological relationship to graphs. You can view a communications network as a series of transmission paths that are used to interconnect different devices to include terminals, multiplexers, routers, concentrators, port selectors, and similar equipment.

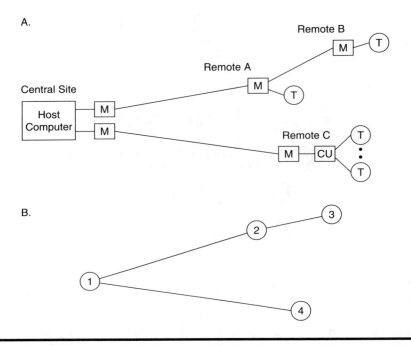

Figure 5.1 Network and Graph Relationship
A communications network can be represented as a graph by considering
each location as a node and circuits between nodes are represented as links:
(A) simple communications network (M — modem, CU — control unit, T —
terminal) and (B) network in graph format.

5.1.1 Links and Nodes

If you consider that transmission paths as branches with equipment
clustered at a common location are used to represent a node, you can
redesign a communications network in the form of a graph consisting of
branches that are formally referred to as links and nodes, which are
connected to one another by one or more links. To illustrate this concept,
consider the network illustrated at the top portion of Figure 5.1 and its
graphical model illustrated at the bottom.

In converting the network schematic at the top of Figure 5.1 to its
graphical format, note that the host computer and the two modems at the
central site location are considered as an entity and denoted as Node 1.
The terminal and modem at sites Remote A and Remote B are also
considered as an entity for Node 2 and Node 3. Finally, the modem,
control unit, and set of terminals at Remote C are considered to represent
Node 4, when the communications network is redrawn as a graph. Also
note that if you replaced the modems in the top portion of Figure 5.1 by

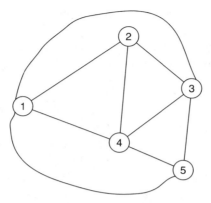

Figure 5.2   Five-Node Graph

DSUs and the control unit and terminals by a router and LAN, the WAN graph shown in the lower portion of the referenced illustration would remain unchanged. Thus, the use of graphs is a powerful mechanism to determine equipment locations regardless of the type of communications equipment used in a network.

Now that we have briefly examined the relationship of graph theory to communications networks, let us focus our attention upon the properties of graphs that will enable us to consider the use of a connection matrix that can be developed to describe the paths between nodes and the relationship between the links connecting each node. Once this is accomplished, we will focus our attention upon the typical network design problem concerning where to locate data concentration equipment and the use of graph theory to solve this problem. Thus, the intent of this chapter is to provide practical information in the form of graph theory knowledge that can be applied to solve a common network design problem. Later in this book, we will use the knowledge of graph theory presented in this chapter to solve another common network design problem — obtaining an optimum route for a multidrop line.

5.1.2   Graph Properties

In examining the network represented in graph format in the lower portion of Figure 5.1, it should be recognized that prior organizational requirements and constraints resulted in the original network design that the graph represents. Rather than considering requirements and constraints at this time, let us start anew by examining the representation of a potential network in which four remote locations are to be connected to a host computer. Figure 5.2 illustrates a graph that contains five nodes and which will be used to both discuss the properties of graphs, as well as develop and examine the properties of a connection matrix resulting from the topology of the graph.

The five-node graph in Figure 5.2 can represent a central computer site and four remote locations or it can represent five remote locations clustered within one geographical area.

The versatility associated with using graphs to represent networks can be envisioned by considering a few of the possible network configurations Figure 5.2 could represent. This diagram could represent a central computer site and four remote sites that must communicate directly with the central site or it could represent the possible routes allowed for the construction of a single multidrop line connecting each remote site to a central computer. As another possibility, the diagram could represent five remote locations clustered within a geographical area that must communicate with a distant computer system, such as five cities in a state east of Mississippi that must communicate with a computer located in Denver.

The five-node graph illustrated in Figure 5.2 contains nine links, where each link represents a line or branch connecting two nodes. As such, each link can be defined by a set of node pairs, such as (1.2), (1.3), (1.4), (1.5), and so on. If we assume that the flow of communications represented by a graph is bidirectional we do not have to distinguish between link (i,j) and link (j,I) where i and j represent node numbers. Thus, we can define link (i,j) = link (j,I) and will do so. Furthermore, we can define the graph representing a set of N nodes and L links as $G(N,L)$.

5.1.2.1 Network Subdivisions

If we subdivide a network, we can represent the subdivision as a subgraph of $G(N,L)$. As an example of network subdivision, assume the network illustrated at the top of Figure 5.1 was subdivided into the sets of node pairs (1,4) and (1,2), (2,3), and (3,5). Then, $G_1(N_1,L_1)$ represents the subgraph of the network whose node pair is (1,4). Also, $G_2(N_2,L_2)$ represents the subgraph whose set of node pairs is (1,2), (2,3), (3,5) and whose links connect Node 1 to Node 2, Node 2 to Node 3, and Node 3 to Node 5, then the union of G_1 and G_2 (represented mathematically as $G_1 \cup G_2$) represents the undivided or original network. Thus, $G_1 \cup G_2$ is the graph whose set of nodes is $N_1 \cup N_2$ and whose set of links is $L_1 \cup L_2$.

5.1.2.2 Routes

If i and j are distinct nodes in the graph $G(N,L)$, we can define the term route as an ordered list of links (i,i_1), $(i_1,i_2),\dots(I_n,j)$, such that i appears only at the end of the first link and j appears only at the end of the last link. With the exception of nodes i and j, all other nodes in the route will appear exactly twice, as they represent an entrance into a node from one link as well as an exit of the node by a link homing on another node. Based upon the preceding route properties, it will not contain a loop nor will it represent the retracing of a link.

5.1.2.3 Cycles and Trees

Two additional properties of graphs that warrant attention are cycles and trees. A cycle is a route that has the same starting and ending nodes. Thus, the route (1,2), (2,4), (4,5) in Figure 5.2 could be converted into the cycle (1,2), (2,4), (4,5), (5,1) by using link (5,1) to return to Node 1. Note that link (5,1) could be replaced by link (5,3) and link (3,1) to obtain a different cycle that also returns to Node 1. A tree is a collection of links that connect all nodes in a graph and whose links contain no cycles.

Although a comparison between a route and a tree may appear trivial, in actuality the difference between the two is both distinct and has a key applicability to one specific area of network design. Unlike a route that may or may not connect all nodes in a graph, a tree will connect all nodes. This means that an analysis of the length of different trees that can be constructed from a common graph can be used, as an example, to determine the optimum route for a multidrop line that is used to interconnect network locations represented as nodes on a graph.

5.1.3 The Basic Connection Matrix

We can describe a graph $G(N,L)$ that contains N nodes and L links by an N-by-N matrix. In this matrix we can assign the value of each element based upon whether or not a link connects Node i to Node j. Thus, if X is the connection matrix, we can describe the value of each element X_{ij} as follows:

$X_{ij} = 0$ if no link connects i to j
$X_{ij} = 1$ if there is a link between i and j

In the preceding definition, we will use the variable i to represent the ith row of the connection matrix, and the variable j will be used to represent the jth column of the connection matrix. Figure 5.3 represents the connection matrix for the graph illustrated in Figure 5.2. In a basic connection matrix, $X_{ij} = 1$ when a link connects nodes i to j; otherwise $X_{ij} = 0$.

5.1.3.1 Considering Graph Weights

Although a basic connection matrix illustrates the relationship between links and nodes, that relationship is expressed as a binary relationship.

$$X = \begin{bmatrix} 0 & 1 & 1 & 1 & 1 \\ 1 & 0 & 1 & 1 & 0 \\ 1 & 1 & 0 & 1 & 1 \\ 1 & 1 & 1 & 0 & 1 \\ 1 & 0 & 1 & 1 & 0 \end{bmatrix}$$

Figure 5.3   Basic Connection Matrix

Table 5.1   Link Distances

Link	Distance (Miles)
(1,2)	9
(1,3)	11
(1,4)	8
(1,5)	13
(2,3)	7
(2,4)	14
(2,5)	12
(3,4)	8
(3,5)	6
(4,5)	6

In reality, we normally wish to consider the assignment of values to links where such values can represent the length of a link, its cost, transmission capacity, or some similar value as long as the values for all links are expressed in the same term. The assignment of values to links results in a real number known as a weight being placed on, above, or below each link on a graph.

If we denote the weight of a link between nodes i and j as W_{ij}, then we can assume that $W_{ij} = W_{ji}$, because we will not distinguish between link (i,j) and link (j,i). Similarly, if there is no link between nodes i and j, W_{ij} will be assigned the value 0.

To illustrate the use of graph weights, let us consider a five-node graph in which each node is connected by a link to another node. Let us assume that the link distances between nodes are as indicated in Table 5.1.

The distances listed in Table 5.1 can be used to construct a weighted connection matrix. This matrix is illustrated in Figure 5.4 and could represent a geographical area consisting of a city and suburbs in which five offices are located. In this instance, a typical network design problem

$$X = \begin{bmatrix} 0 & 9 & 11 & 8 & 13 \\ 9 & 0 & 7 & 14 & 12 \\ 11 & 7 & 0 & 8 & 6 \\ 8 & 14 & 8 & 0 & 6 \\ 13 & 12 & 6 & 6 & 0 \end{bmatrix}$$

Figure 5.4   Weighted Connection Matrix

you may encounter is to determine the optimum location for the installation of a control unit among the five offices. In this type of problem, you would consider each location as a potential site to install the control unit. Because a terminal in each of the other offices would be connected to the control unit by the use of a leased line, you would want to select the control unit installation location to minimize the distance to all other offices.

In a weighted connection matrix, each element is assigned a common value that can represent line cost, distance, or traffic handling capacity.

As previously discussed, a weighted connection matrix can be used to represent all possible connections between a number of offices within a geographical area. If we assume that one terminal device is to be located at each office and we wish to connect those terminals to a common control unit, multiplexer, or router, the question arises of where to place the communications device. In effect, this problem is a minimum route distance problem that can be solved by the manipulation of the elements of the previously developed weighted connection matrix.

In the next section of this chapter, we will use the information concerning graph theory presented in this section to solve several typical data concentration equipment location problems. First, we will examine the placement of one terminal per location. Next, we will vary the number of terminals we will develop and modify a Basic language program you can use to automate the equipment location process.

5.2   EQUIPMENT LOCATION TECHNIQUES

One of the most common problems associated with the use of data concentration equipment is determining an optimum location to install such equipment. When the terminal population that will be serviced by data concentration equipment is clustered within a building or small geographical area, including a city and its suburbs, the problem of equipment location is typically reduced to selecting a convenient site that has available space, electrical power, and personnel available to provide access

Table 5.2   Distances between Terminal Locations

Terminal Locations	Distance (Miles)
1–2	90
1–3	110
1–4	80
1–5	130
2–3	70
2–4	140
2–5	120
3–4	80
3–5	60
4–5	60

to installers and repair personnel if, at a later date, servicing or upgrading equipment becomes necessary. In this type of situation, little thought is given nor usually required to consider the cost of communications between each terminal device and potential data concentration locations to determine an optimum location based upon the cost of communications.

5.2.1   Examining Distributed Terminals

In place of the situation where terminal devices are clustered, let us consider the effect upon the distribution of terminals over a wider geographical area. For illustrative purposes, let us assume there are five locations at which one terminal will be installed. Let us further assume that the distances between terminal locations are as indicated in Table 5.2.

In examining the distances between terminal locations, it is not apparent which location would be optimum with respect to the distance to all other locations for the installation of a multiplexer, router, or data concentrator. If we consider each terminal location as a node, we can consider the distances between nodes as a weighted link for the development of a weighted graph. Then we can apply our prior discussion of graph theory to solve this problem.

Using the data contained in Table 5.2, we can develop a weighted graph as illustrated in Figure 5.5. Note that this graph uses nodes to represent terminal locations and the distances between locations are expressed as link weights. This graph represents the geographical distribution of terminals in miles from one another.

We can use the distances between terminal locations contained in Table 5.2 or the graphical representation of the distributed terminal locations illustrated in Figure 5.5 to create a weighted connection matrix. This

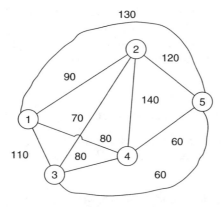

Figure 5.5 Graphical Representation of Distributed Terminal Locations

$$X = \begin{bmatrix} 0 & 90 & 110 & 80 & 120 \\ 90 & 0 & 70 & 140 & 120 \\ 110 & 70 & 0 & 80 & 60 \\ 80 & 140 & 80 & 0 & 60 \\ 130 & 120 & 60 & 60 & 0 \end{bmatrix}$$

Figure 5.6 Weighted Connection Matrix

matrix is illustrated in Figure 5.6 and allows us to easily consider each location as a potential site for the installation of data concentration equipment. In Figure 5.6, the weighted connection matrix uses the assignment of mileage to each element.

5.2.2 Using a Weighted Connection Matrix

A weighted connection matrix can be used to represent all possible connections between terminal locations within a geographical area. When we desire to find an optimum location to place data concentration equipment, in effect, we are attempting to solve a minimum route distance problem. This problem is easily solved by the manipulation of the elements contained in the previously developed weighted connection matrix.

Although we can easily analyze the elements in the weighted connection matrix contained in Figure 5.6, suppose there were 20, 30, 50, or 100 offices within the geographical area we had to consider. In such situations, it would be preferable to develop a computer program to analyze the connection matrix to determine an optimum location to install data concentration equipment.

Due to the use of computer programs facilitating the solution of data concentration equipment location problems, let us examine the development and use of a computer program for analyzing the 5-by-5 element connection matrix and which can be expanded to solve more complex problems.

5.2.3 Automating the Location Process

Table 5.3 contains a Basic language program listing that, when executed, computes the route distance from each node in the graph illustrated in

Table 5.3   Node Location Program Listing

```
REM PROGRAM TO COMPUTE NODE LOCATION
CLS
DIM X(5, 5), DISTANCE(5)
FOR I = 1 TO 5
FOR J = 1 TO 5
READ X(I, J)
NEXT J
NEXT I
DATA 0,90,110,80,130
DATA 90,0,70,140,120
DATA 110,70,0,80,60
DATA 80,140,80,0,60
DATA 130,120,60,60,0
REM FIND MINIMUM ROUTE DISTANCE
FOR I = 1 TO 5
DISTANCE(I) = 0
FOR J = 1 TO 5
DISTANCE(I) = DISTANCE(I) + X(I, J)
NEXT J
NEXT I
REM PRINT RESULTS
FOR J = 1 TO 5
LOW = DISTANCE(1)
K = 1
FOR I = 1 TO 5
IF LOW < DISTANCE(I) GOTO 1
LOW = DISTANCE(I)
K = I
1 NEXT I
PRINT "NODE "; K; " ROUTE DISTANCE ="; LOW
DISTANCE(K) = 9999
NEXT J
```

Figure 5.5. In effect, this program can be used to obtain the information necessary to select an optimum node location from a five-node graph and can be easily modified to operate on graphs containing additional nodes.

The DIM (dimension) statement allocates space for a two-dimensional 5-by-5 (25) element array labeled X, which will represent the connection matrix and a one-dimensional five-element array labeled DISTANCE. The first pair of FOR-NEXT loop statements initializes the element values of the connection matrix as each time a loop occurs the READ statement assigns a value in a DATA statement to X_{ij} based upon the values of I and J in the FOR-NEXT loop. Note that the values in the DATA statements correspond to the row values of the connection matrix previously listed in Figure 5.6.

The second pair of FOR-NEXT statement loops in the program listing first initialize each element of the one-dimensional DISTANCE array to zero. Next, the route distance from Node i to all other nodes is computed as J varies from 1 to 5. In effect, the preceding operations sum the row values of the connections matrix.

The last portion of the program listed in Table 5.3 prints the results based upon the values of the elements of the DISTANCE array. First, the variable LOW, which represents the lowest number or route distance, is set to the value of the first element in the DISTANCE array. Next, as the variable I varies from 1 to 5, which correspond to all elements in the DISTANCE array, the value of LOW is compared to the value of each element of the DISTANCE array. If the value of LOW is not less than the value of the appropriate DISTANCE element, the value of the DISTANCE element is assigned to LOW and the variable I, which indicates the node associated with the DISTANCE element, is assigned to the variable K. If the value of LOW is less than the value of the appropriate DISTANCE element, a branch to the statement labeled with 1 occurs, in effect, bypassing the assignment of the DISTANCE element to LOW and the value of K to I. Once the I loop is completed, K represents the node number that has the lowest or shortest route distance and the variable LOW contains the value of the route distance. After the PRINT line is executed, the Kth element of the DISTANCE array is assigned the value 9999. In effect, this operation precludes the selection of this node a second time when J is incremented by one and the J loop is again executed. Thus, the nested FOR-NEXT J and I loops result in the printing of the route distance by node in ascending distance. Table 5.4 illustrates the execution of the program based upon the use of the connection matrix illustrated in Figure 5.6.

As illustrated in Table 5.4, Node 3 represents the optimum location to place a data concentration device based upon the previously described problem. In this example, the placement of data concentration equipment at Node 3 results in a reduction of 40 miles in comparison to the selection

Table 5.4   Execution of Node Location Program

NODE 3	ROUTE DISTANCE = 320
NODE 4	ROUTE DISTANCE = 360
NODE 5	ROUTE DISTANCE = 370
NODE 1	ROUTE DISTANCE = 410
NODE 2	ROUTE DISTANCE = 420

of Node 5, 90 miles over the selection of Node 1, and 100 miles in comparison to selecting Node 2. Because the monthly cost of leased communications facilities is based upon mileage, this solution also represents an optimum economic location.

5.2.4   Extending the Node Location Problem

In solving the previous node location problem, it was assumed that only one terminal would be installed at each location and that all terminals would communicate over leased lines. These two assumptions enabled the use of a symmetrical connection matrix, because the cost of a leased line from A to B is the same as a line from B to A and limiting one terminal per location ensures that the weight of link X_{ij} equals the weight of link X_{ji}.

Suppose there are some locations that will have more than one terminal device installed. What effect does this situation have on the node location problem and the symmetry of the connection matrix? To illustrate the effect of increasing the terminal population, let us first consider dial or switched network access from each terminal to a data concentration device located at a node. Next, we will examine the use of leased lines to support a nonuniform distribution of terminals. Then, we can use the previously developed networking information to examine the effect of using a mixture of switched network and leased line facilities to service a terminal population within a geographical area by common data concentration equipment.

5.2.5   Switched Network Utilization

In place of a uniform distribution of one terminal per node, let us assume a variable distribution of terminals. In addition, let us further assume that the PSTN will be used as the data transportation medium to the location where a multiplexer, router, or another type of data concentration equipment will be located.

To determine the cost associated with the use of the switched network from one location to another would require the development of a complex table in which the cost of a call from each location to every other location would be listed. Fortunately, the tariff structure for PSTN calls is commonly based upon mileage bands, with calls between cities that fall into the same mileage band distance billed at the same rate. Due to this, in many instances, it becomes possible to simplify the construction and application of a cost table. If we assume the terminal locations in the geographical area that we are examining are located within the same switched network mileage band, we can use the same cost elements in computing the cost per call from one location to all other locations. That is, we will assume the cost of the first minute and each succeeding minute is the same from Location 1 dialing Location 3 as it is for Location 1 dialing Location 5 and Location 2 dialing Location 1 and so on. Otherwise, we would have to determine the mileage band for calls from each node to all other nodes.

Assuming the mileage bands for all locations calling all other locations are equivalent, our cost table can be easily constructed. Table 5.5 illustrates the construction of a Switched Network Cost Table. In constructing this table, you would normally include some additional information that was omitted for simplicity, such as the average call duration in minutes and the cost per minute. These two missing columns of data were replaced by the Average Cost per Call column in the referenced table.

The row entries in the Total Cost per Day column in Table 5.5 were computed by multiplying the entries in Column 2 by the entries in Column 3 by the Cost per Call entry in Column 4. Next, it was assumed that the organization was actively transmitting data an average of 22 days per month. Thus, the Monthly Cost column was completed by multiplying the Total Cost per Day column entries by 22.

The entries in Column 6 of Table 5.5 can be used to develop a nonsymmetrical connection matrix where X_{ij} may or may not be equal to X_{ji}. This is because a nonuniform distribution of terminals, a nonuniform number of calls per day per terminal, and a nonuniform cost per call either by themselves or collectively contribute to the weight of X_{ij} being less than, equal to, or greater than the weight of link X_{ji}, where the link weight from i to j is the monthly cost of communications from Location i to Location j.

Table 5.5 Switched Network Cost Table

Location	Number of Terminals	Calls/Day per Terminal	Average Cost per Call	Total Cost per Day ($)	Monthly Cost ($)
1	1	2	1.00	2.00	44.00
2	2	1	2.00	4.00	88.00
3	2	3	1.50	9.00	198.00
4	1	4	2.00	8.00	176.00
5	3	2	2.00	12.00	264.00

$$X = \begin{bmatrix} 0 & 44 & 44 & 44 & 44 \\ 88 & 0 & 88 & 88 & 88 \\ 198 & 198 & 0 & 198 & 198 \\ 176 & 176 & 176 & 0 & 176 \\ 264 & 264 & 264 & 264 & 0 \end{bmatrix}$$

Figure 5.7   Switched Network Utilization Connection Matrix

Using Column 6 in Table 5.5, we can create the connection matrix illustrated in Figure 5.7. When the number of terminals or call duration per terminal varies, a weighted connection matrix will not be symmetrical. In creating the connection matrix, the monthly cost of communications from Location 1 ($44) is assumed to be the cost from that location to all other locations. Similarly, the cost of communications from Location 2 to all other locations was assumed to be $88. Note that the differences in the number of terminals, calls per day, and cost per call between Location 1 and Location 2 ensured $X_{12} \neq X_{21}$.

In attempting to use the connection matrix illustrated in Figure 5.7 as a basis for use in the previously described Basic NODE analysis program, your first inclination might be to simply substitute the entries in the connection matrix into the DATA statement values in the program. If you did so, your DATA statements would appear as indicated in Table 5.6.

Using the previously discussed modifications, the execution of the program would result in the computation of route distances, which now represent the monthly cost of communications, as indicated in Table 5.7. However, the question arises — are these distances or costs correct?

To verify the correctness of the program execution, let us examine a route distance. From Figure 5.7, the route distance for Node 1 is 176, which represents the distance or cost from Location 1 to all other locations. Because the connection matrix is not symmetrical as it does not represent an even bidirectional weight, what the program computed represents is the reverse direction of the problem, that is, the placement of data concentration equipment at each location requires the computation of the total route distance or cost from all other node locations to that location.

Table 5.6   Modified DATA Statements

```
DATA 0,44,44,44,44
DATA 88,0,88,88,88
DATA 198, 198, 0, 198, 198
DATA 176, 176, 176, 0, 176
DATA 264, 264, 264, 264, 0
```

Table 5.7 Executing of Node Location Program Using Revised Data

NODE 1	ROUTE DISTANCE = 176
NODE 2	ROUTE DISTANCE = 352
NODE 4	ROUTE DISTANCE = 704
NODE 3	ROUTE DISTANCE = 792
NODE 5	ROUTE DISTANCE = 1056

Table 5.8 Changing the Program to Sum by Column

```
REM FIND MINIMUM ROUTE DISTANCE
FOR J = 1 TO 5
DISTANCE (J) = 0
FOR I = 1 TO 5
DISTANCE (J) = DISTANCE (J) + X (I, J)
NEXT I
NEXT J
```

Table 5.9 Execution of Revised Node Location Program

NODE 5	ROUTE DISTANCE = 506
NODE 3	ROUTE DISTANCE = 572
NODE 4	ROUTE DISTANCE = 594
NODE 2	ROUTE DISTANCE = 682
NODE 1	ROUTE DISTANCE = 726

Thus, the program must be modified to sum the connection element values by column instead of by row. Table 5.8 illustrates the program segment change in which the I and J FOR-NEXT loops were reversed to enable the summation of element values by column. Table 5.9 illustrates the execution of the program based upon the previously described program modifications.

5.2.6 Other Program Modifications

To simplify the computation of the cost of leased lines based upon their distance between rate centers, AT&T developed a vertical and horizontal (VH) Coordinate Grid System that overlays all of the continental United States and a large portion of Canada at an approximate 30 to 45 degree shift from a true north to south orientation. The VH Coordinate Grid System varies both V and H from 0 to 10,000, with H varying from 0 to

10,000 going from right to left on the *VH* map, and *V* varying from 0 to 10,000 going from top to bottom on the map. This results in some upper Maine locations having *VH* coordinates of 3000, 0 and upper Washington state locations having *VH* coordinates of approximately 8000, 10,000. At the lowest *V* level, the Florida Keys and some locations in southern Texas have a *V* coordinate of approximately 10,000 and the *H* coordinate value varies from approximately 0 at the eastern edge of the Florida Keys to approximately 4000 near the tip of southern Texas. Figure 5.8 illustrates the AT&T *VH* Coordinate Grid System overlaid on the eastern and midwestern portions of the United States.

The *VH* Coordinate Grid System is used by most communications carriers, as well as a number of independent firms, as the foundation for developing programs that compute the location to place a data concentration node within a geographical area, the path that represents a minimum route multidrop circuit, and other network design functions. Because each rate center corresponds to an area code and three-digit telephone prefix, a number of firms sell software that allows users to enter those six digits that represent office locations to obtain their *VH* coordinate location. Other firms sell tables listing *VH* coordinates of thousands of cities in the United States and Canada, and some vendors market a database of *VH* coordinates that can be incorporated into user-developed software. Assuming you have access to one or more methods to obtain *VH* coordinates for your organizational locations, let us examine how the previously described Basic language program could be modified to automatically compute link distances.

The grid formed to define the *VH* boundaries between two locations represents a two-dimensional plane. Due to this, a slightly modified version of the Pythagorean Theorem can be used to calculate the distance between pairs of *VH* coordinates representing two locations. In the Pythagorean Theorem, the hypotenuse of a right triangle is shown to be equal to the square root of the sum of the squares of the other two sides of the triangle. Thus, if C is the hypotenuse and A and B are the other two sides of the triangle, the length of the hypotenuse becomes:

$$C = \left(A^2 + B^2\right)^{1/2}$$

In the *VH* Coordinate Grid System, you can envision *A* and *B* to be the difference between pairs of *VH* coordinates. Thus, you can replace *A* by $V_1 - V_2$ and *B* by $H_1 - H_2$ where the subscripts, 1 and 2, represent Location 1 and Location 2. Finally, to convert *VH* coordinate points to mileage, you must divide the resulting sum of the squares by 10 prior to taking the square root, thus, the formula for calculating the distance (*D*) in miles between two locations expressed as *VH* coordinates becomes:

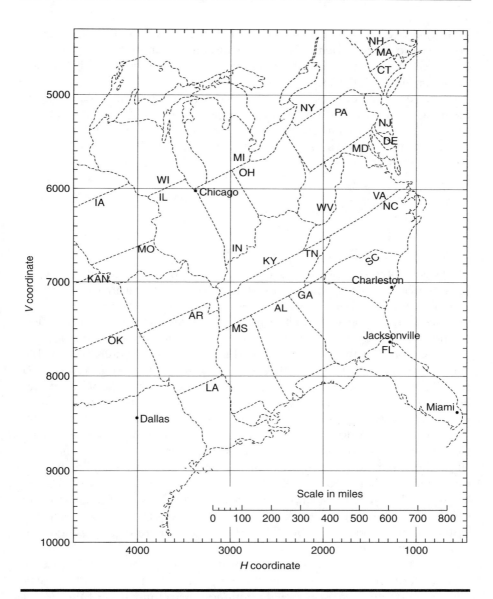

Figure 5.8   Portion of the AT&T *VH* Coordinate Grid System

$$D = INT \left(\left(\frac{\left(V1 - V2\right)^2 + \left(H1 - H2\right)^2}{10} \right)^{1/2} + 0.5 \right)$$

In the preceding equation, 0.5 is added to the result of the computation prior to the integer being taken, because a communications carrier is permitted by tariff to round the computed mileage to the next higher mile in performing its cost calculations.

As an example of the use of the *VH* coordinate system, let us compute the mileage between Denver and the geographical center of Georgia located in the famous town of Macon. The *VH* coordinates of Denver are 7501, 5899 and Macon's *VH* coordinates are 7364, 1865. Thus, the distance between those two locations is:

$$D = INT\left(\left(\frac{\left(7501 - 7364\right)^2 + \left(5899 - 1865\right)^2}{10}\right)^{1/2} + 0.5\right) = 1277 \text{ miles}$$

Based upon the preceding description of the *VH* Coordinate Grid System, you can easily modify the Basic program listed in Figure 5.5 to compute distances between nodes, rather than work with predefined link distances. Here, you would either input *VH* coordinates for each location or the area code and three-digit prefix of the telephone exchange for each location. For the latter, you could integrate a commercially available database into your program that would allow you to retrieve the *VH* coordinates for a location based upon the area code and telephone prefix of the location. Once the *VH* coordinates for each node are determined, you could use the previously noted equation to determine the link distances between nodes. Thereafter, the program would operate as previously described to compute the total route distance from each node to all of the other nodes in the network. Finally, for networks that have a large number of nodes, you could modify the program to sort and list the route distances in ascending order by node. This would allow users to easily consider alternate data concentration equipment locations if a better location is not suitable for installing the required equipment for some reason.

Now that we have examined the properties of graphs and the use of a connection matrix to determine the optimum location to place data concentration equipment, we will continue our examination of the use of graph theory to solve another common network design problem — where to route a multidrop line to minimize its distance. Thus, in Chapter 6, we will conclude our application of graph theory to network design by examining its use to determine an optimum multidrop line route.

6

MULTIDROP LINE ROUTING TECHNIQUES

One of the frequent problems encountered by organizations is determining an economical route for the path of a multidrop circuit. This type of circuit is commonly used to interconnect two or more locations that must be serviced by a mainframe computer port. Although there are several commercial services that you can subscribe to, as well as a free service offered by some communications carriers to obtain a routing analysis, in many situations, this analysis can be conducted internally within the organization. Doing so not only saves time, but may also eliminate some potential problems that can occur if one relies upon programs that do not consider whether the resulting number of drops on a circuit can support the data traffic while providing a desired level of performance.

In this chapter, we will examine the use of several algorithms that can be employed to minimize the routing distance and resulting cost of a multidrop circuit. Because there is a finite limit to the number of drops a multidrop circuit can support, we will also investigate a method that will enable users to estimate the worst case and average terminal response times as the number of drops increase. Then if the response time exceeds the design goal of the organization, the network manager can consider removing one or more drops and placing them on a different multidrop circuit. Due to the applicability of graph theory for solving multidrop routing problems, we will use information previously presented in this book (see Chapter 5) concerning that area of mathematics as a foundation for solving additional communications related problems.

6.1 MULTIDROP ROUTING ALGORITHMS

In this section, we will examine the operation and use of a popular algorithm that can be employed to develop a multidrop network structure.

This algorithm, which is known as the Prim algorithm in recognition of its developer, was publicized in 1957 in the *Bell System Technical Journal*. Because this algorithm results in the development of a minimum spanning tree (MST) in which the link distances between nodes are minimized, the resulting tree structure obtained by the use of this algorithm is known as a minimum spanning tree and the algorithm is commonly referred to as an MST algorithm. Although this algorithm results in the development of an optimum network structure, it does not consider such constraints as traffic flow or response time. Thus, later in this chapter we will examine methods by which such constraints can be considered.

6.1.1 The MST Technique

When the total number of drops to be serviced does not exceed the capacity of software operating on a computer connected to a multidrop line, the MST technique can be used. This technique results in the most efficient routing of a multidrop line by the use of a tree architecture, which is used to connect all nodes with as few branches as possible. When applied to a data communications network, the MST technique results in the selection of a multidrop line whose drops are interconnected by branches or line segments, which minimize the total distance of the line connecting all drops. Because the distance of a circuit is normally proportional to its cost, this technique results in a multidrop line that is also cost optimized. To better understand the procedure used in applying the MST technique, let us first examine an example of its use.

The upper portion (A) of Figure 6.1 illustrates the location of a mainframe computer with respect to four remote locations that require a data communications connection to the computer. Assuming that remote terminal usage requires a dedicated connection to the computer, such as busy travel agency offices might require to their corporate computer, an initial network configuration might require the direct connection of each location to the computer by separate leased lines. The lower portion (B) of Figure 6.1 illustrates this network approach.

When separate leased lines are used to connect each terminal location to the mainframe, a portion of many line segments can be seen to run in parallel. Thus, from a visual perspective, it is apparent that the overall distance of one circuit linking all locations to the computer will be less than the total distance of individual circuits. Other factors that can reduce the cost of a composite multidrop circuit in comparison to separate leased lines includes differences in the number of computer ports and modems or data service units required between the use of a multidrop line and individual leased lines.

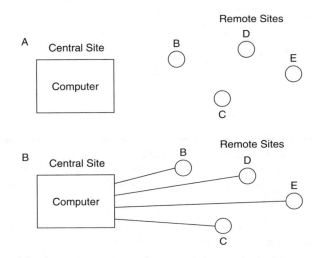

Figure 6.1   Terminal Locations to Be Serviced
(A) Terminal locations with respect to central site computer. (B) Individual site connection to computer.

A multidrop line requires the use of one computer port with a common modem or service unit servicing each of the drops connected to the port. A total of n + 1 modems or service units are thus required to service *n* drops on a multidrop circuit. In comparison, separate point-to-point leased lines would require one computer port per line as well as n + 2 modems or service units, where *n* equals the number of required point-to-point lines.

6.1.1.1   Applying Prim's Algorithm

To apply Prim's algorithm, which is also known as an MST algorithm, we must first convert the diagram showing terminal locations to be serviced in Figure 6.1 into a graph. Figure 6.2 illustrates the result of this conversion. Note that the central site computer has been replaced by Node 1 and the remote sites previously labeled B, C, D, and E were replaced by nodes labeled 2 through 5.

Now that we have established a graph representing the network, the next step to be performed prior to applying Prim's algorithm is to assign link weights to the graph. Here, the assignment of link weights will be based upon the distance between nodes because the cost of a leased line is normally proportional to its length. Due to the complexity of tariffs, this may not always be correct, and you may wish to assign values to link weights based upon the actual cost of leased lines between each node.

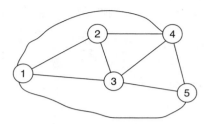

Figure 6.2 Conversion of Network Locations into a Graph
This figure illustrates the conversion of the network locations contained in Figure 6.1 into a graph. Note that the central site computer was assumed to be Node 1.

For illustrative purposes, let us assume the distances between the nodes previously illustrated in Figure 6.2 correspond to the entries in Table 6.1. We can then use those distances between nodes as the link weight assignments and revise Figure 6.2. Figure 6.3 illustrates this revision in which each link has been assigned a link weight.

Prim's algorithm lets us start the construction of an MST by selecting any node in a graph. Once a node is selected, you then construct a two-node subgraph by connecting the first node to the nearest node, in effect, selecting the minimum link distance between the first node and all other nodes in the graph. Next, you will expand the two-node subgraph into a three-node subgraph by connecting one of the nodes in the two-node subgraph to the nearest node not contained in the subgraph. This process is repeated until all of the nodes are connected to one another.

To illustrate the use of Prim's algorithm, let us apply it to the graph containing link weights representing the mileage between nodes previously

Table 6.1   Node Distance Relationship

Link	Node Distance
1,2	30
1,3	60
1,4	60
1,5	90
2,3	50
2,4	40
3,4	20
3,5	50
4,5	30

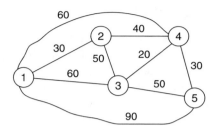

Figure 6.3   Assignment of Link Weights
This figure illustrates the assignment of node distances to the network graph
contained in Figure 6.2.

Table 6.2   Constructing the MST

Step	Nodes in Subgraph	Link Addition
1	1	(1,2)
2	1,2	(2,4)
3	1,2,4	(4,3)
4	1,2,3,4	(4,5)
5	1,2,3,4,5	------

illustrated in Figure 6.3. Table 6.2 lists the steps involved in developing
an MST based upon the creation and expansion of a subgraph and the
resulting link addition each time the subgraph is expanded. In this example, the construction of the MST commenced by selecting Node 1 as the
starting node, however, you could obtain the same tree structure by
selecting any other node in the graph as the starting node. Figure 6.4
illustrates the resulting MST based upon Prim's algorithm. Note that this
tree has a link weight of 120.

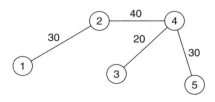

Figure 6.4   Resulting MST
This MST has a link weight of 120.

6.1.1.2 Considering Fan-Out

In examining Figure 6.4, one item of special interest that warrants attention is the dual links that home on Node 4. In a multidrop network configuration, a modem or service unit installed at Node 4 would have to contain a fan-out feature to support multiple links. Otherwise, the use of conventional multidrop modems or DSUs would preclude the use of the MST configuration previously illustrated in Figure 6.4. In such situations where you cannot or prefer not to use modems or service units with a fan-out feature, you may want to use a modified MST technique that results in a minimum tree length with only one link routed from a node to any other node.

6.1.2 Modified MST Technique

In the modified MST technique, you should select the farthest location from the node connected to the computer first. In our example, this would be Node 5. Next, form subgraphs similar to the manner previously described, adding the following constraints:

- If two links currently connect a node in a subgraph, exclude the use of that node for a link when you expand the subgraph by selecting a link to the nearest nonconnected node.
- Home on the destination link last.

Based upon the preceding constraints, Table 6.3 lists the steps followed in developing a modified MST. Figure 6.5 illustrates the resulting tree. Note that the link weight of the tree has increased to 130, however, it does not require the use of modems or service units with a fan-out feature.

In examining Table 6.3, let us review each of the steps to illustrate the procedure required to avoid a fan-out situation. In Step 1, the most distant node, Node 5 in this example, is selected. From Node 5, the nearest node is Node 4, hence, link (5,4) is added for the formation of our

Table 6.3   Constructing a Modified MST

Step	Nodes in Subgraph	Link Addition
1	5	(5,4)
2	5,4	(4,3)
3	5,4,3	(3,2)
4	5,4,3,2	(2,1)
5	5,4,3,2,1	------

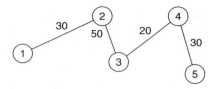

Figure 6.5   Revised Tree
The revised tree contains a maximum of two links connecting a node, elimi-
nating the necessity of using modems with a fan-out feature when constructing the network configured by the graph.

multidrop line. Because the distance from Node 4 to other nonconnected
nodes are less than the distance from Node 5 to other nonconnected
nodes, we will resume our work at Node 4. From that node we would
then select Node 3 as the nearest node and add the link (4,3) to the
formation of our multidrop line. Next, in examining the distance from
Node 3, Node 4, and Node 5 to the remaining unconnected nodes, we
would select Node 4 to work with if we were developing a conventional
MST structure. However, if we work with Node 4, it would violate our
previously stated constraint designed to prevent a fan-out situation, as
Node 4 already has two links connected to it. Thus, to prevent a fan-out
situation from occurring, we should select Node 3 to work with.

By selecting Node 3 instead of Node 4 to use as a base for forming
a link to Node 2, we eliminate a fan-out from occurring. However, we
also replace a link (4,2) whose weight is 40 by a link (3,2) whose weight
is 50, resulting in an increased link weight during the formation of the
modified MST. Lastly, we use Node 2 as a base and connect it to Node
1 to complete our modified MST whose structure does not include a
fan-out.

6.1.3   Using a Connection Matrix

To illustrate the use of a connection matrix, let us focus our attention
upon the initial graph previously illustrated in Figure 6.3. The connection
matrix for that graph is illustrated in Figure 6.6.

We can computerize the development of an MST by first initializing a
two-dimensional matrix using the connection matrix values shown in
Figure 6.6 for each element array value. Once this is accomplished, we
can initialize a one-dimensional array whose element values contain the
nodes in the graph that must be connected to form a tree. Thus, this array
would contain five elements whose values would be 1, 2, 3, 4, and 5.

The computer program developed to determine the MST would first
scan Column 1 of the two-dimensional matrix, searching for the minimum

$$X = \begin{bmatrix} 0 & 30 & 60 & 60 & 90 \\ 30 & 0 & 50 & 40 & 0 \\ 60 & 50 & 0 & 20 & 50 \\ 60 & 40 & 20 & 0 & 30 \\ 90 & 0 & 50 & 30 & 0 \end{bmatrix}$$

Figure 6.6   Connection Matrix for Graph in Table 6.4

distance from Node 1 to another node. In our example, element X_{21} would be selected, because its value is the lowest of all values in Column 1. Because Node 1 and Node 2 are then connected, the elements X_1 and X_2 in the one-dimensional node array whose values were 1 and 2 would be assigned the value zero to denote they are included in a subgraph used to form a tree. Similarly, because $X_{ij} = X_{ji}$, X_{12} in the two-dimensional matrix would be assigned the value zero prior to searching Column 2 in the two-dimensional array for the element with the smallest value. Figure 6.7 illustrates the composition of the node matrix and the connection matrix after the first column in the connection matrix was processed.

The selection of element X_{21} in the connection matrix can also be used as an identifier of the next column in the connection matrix to search. This is because the selection of element X_{21} indicates not only the selection of a link (1,2), but, in addition, that the remaining element values in Column 1 are greater than element values in Column 2. Hence, the development of the MST should commence with Node 2. Thus, Column 2 in the two-dimensional connection matrix is examined next for the minimum element value.

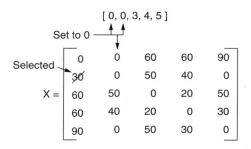

Figure 6.7   Node and Connection Matrix Values
This figure illustrates the composition of the node and connection matrices after the first column in the connection matrix is produced.

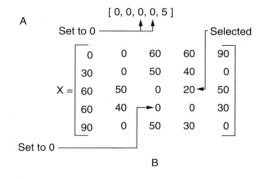

Figure 6.8   Revised Node (A) and Connection Matrix Values (B)

The search of Column 2 in the connection matrix previously illustrated in Figure 6.7 results in the selection of element X_{42} whose value is 40. The selection of that element indicates that link (2,4) is added to the developing MST, as well as the next column to search, in this instance Column 4. Figure 6.8 illustrates the values for the node matrix and the connection matrix at this time. Note that because $X_{ij} = X_{ji}$, X_{34} was assigned the value zero after X_{43} was selected. Also note that X_3 and X_4 in the node matrix were set to zero because the selection of link (4,3) indicates the connection of Node 4 to Node 3.

In examining the revised node matrix illustrated in Figure 6.8, note that only element X_5 whose value is 5 is nonzero. This means that the MST can be completed by connecting Node 5 to a previously connected node. Searching Column 5 in the connection matrix indicates that element X_{45} has the lowest nonzero value, indicating that the selection of link (4,5) is all that remains to complete the development of the MST.

From the preceding description of the manipulation of the values of the elements in the node and connection matrices, you will understand how MST computer programs can be developed. Essentially, the key to such programs is a searching routine that minimizes computer processing time when a large number of nodes results in a large n-by-n matrix whose columns must be searched n times to find the lowest value in each column.

6.2   AUTOMATING THE MST PROCESS

Now that we have discussed the manipulation of the elements in node and connection matrices to develop an MST, let us focus our attention upon a Basic language program that can be used to perform the required computation to solve this type of problem.

6.2.1 Basic Language Program

Table 6.4 contains a Microsoft Quick Basic language program listing that can be executed to solve a modified MST problem that limits each node to one entry and one exit link, in effect, precluding the use of modems with fan-out capability. The entries in this program, which will soon be described, are designed to automate the selection of an MST for the graph illustrated in Figure 6.9. In this illustration, you will note that every location is connected to every other location. If this is not feasible due to such real-world problems as physical access constraints, you can eliminate one or more nodes by assigning a large value to all links connected to a particular node.

The DIM statement in Table 6.4 allocates 5 elements for the node array labeled n and 25 elements for the connection array labeled x, because the program was written to solve a five-node graph problem. The first FOR-NEXT loop initializes the element values of the n array to their node numbers. That is, n_1 is assigned the value 1, n_2 is assigned the value 2, and so on.

The program listing contained in Table 6.4 next contains a dual pair of FOR-NEXT loops using the variable i as the inner loop index and j as the outer loop index. When this pair of FOR-NEXT loops is executed, the READ x(i,j) statement within the loops assigns values in the DATA statement to the two-dimensional connection array based upon rows within a column. This is because i varies from 1 to 5 each time j changes its values. Figure 6.10 illustrates the connection matrix formed by the dual FOR-NEXT loops in the program listing.

The IF statement that follows the READ statement tests the value of each element assigned to the connection matrix. If the value is zero, which denotes a missing link, it is changed to 9999. This change will preclude

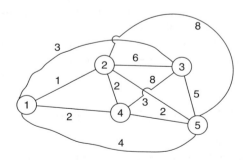

Figure 6.9 Graph to Be Automated
The data statements in the Basic language program listing contained in Table 6.4 describe the link–node relationship of this graph.

Table 6.4   Modified MST Program

```
REM CONNECTION MATRIX FOR MODIFIED MINIMUM SPANNING TREE - NO FANOUT
CLS
PRINT "LINK SELECTION SUMMARY"
DIM n(5), x(5, 5)
REM initialize node matrix element values
FOR i = 1 TO 5
        n(i) = i
        NEXT i
REM initialize connection matrix element values
FOR j = 1 TO 5
FOR i = 1 TO 5
        READ x(i, j)
        IF x(i, j) = 0 THEN x(i, j) = 9999 'make missing links into
large #
        NEXT i, j
        DATA 0,1,3,2,4,1,0,6,2,8,3,6,0,3,5,2,2,3,0,2,4,8,5,2,0
WEIGHT = 0'link weight counter
REM start in column 1
j = 1  'j is the column pointer into connection matrix
1 IF n(j) > 0 GOTO 2'node selected not connected
        FOR k = 1 TO 5
        IF n(k) > 0 THEN j = n(k)'pick first unconnected node
        NEXT k
2       i = 1'pick a legal value in row
4        x = x(i, j)
        IF n(i) > 0 GOTO 3
        i = i + 1
        GOTO 4
3               FOR i = 1 TO 5 'search column j for smallest value
                    IF n(i) = 0 THEN x(i, j) = 9999
                    IF x(i, j) <= x AND n(j) > 0 THEN
                    x = x(i, j)
                    indexi = i   'i index of smallest value in column
                    indexj = j   'j index of smallest value in column
                  END IF
                NEXT i
                IF x = 9999 GOTO 20'when smallest value=9999 we end
                WEIGHT = WEIGHT + x
REM segment selected is link connecting nodes j to i
PRINT "LINK CONNECTS NODES"; indexj; " TO "; indexi; " LINK WEIGHT=";
 x
        n(indexj) = 0'set node element to 0 to represent connection
      REM test if all nodes connected
        FOR i = 1 TO 5
        IF n(i) > 0 THEN 10 'if a node element >0 all nodes not
connected
        NEXT i
        END
10      REM since Xij=Xji, set Xji to high value to preclude its use
        x(indexj, indexi) = 9999
REM search column denoted by indexi which represents node just
connected
        j = indexi
        GOTO 1
20 PRINT "TOTAL LINK WEIGHT =", WEIGHT
```

$$n = \begin{bmatrix} 0 & 1 & 3 & 2 & 4 \\ 1 & 0 & 6 & 2 & 8 \\ 3 & 6 & 0 & 3 & 5 \\ 2 & 2 & 3 & 0 & 2 \\ 4 & 8 & 5 & 2 & 0 \end{bmatrix}$$

Figure 6.10   Connection Matrix Formed by Basic Program
The program listing in Table 6.4 uses the i index for rows and j index for columns.

its use when the program searches each column to locate the lowest value in that column. The variable WEIGHT after the DATA statement is initialized to zero as it will serve as the counter that will hold the cumulative link weights as each link is selected.

Although we can start in any column, for ease of reference, the program starts in Column 1 by assigning the value of 1 to the variable j. Next, the statement labeled with the number 1 for branching purposes uses the column pointer to check the value of the node matrix element associated with that column. If the value is not zero, this indicates that the node was not connected and a branch to the statement labeled 2 occurs. If the node was previously selected, the FOR-NEXT loop using k as the loop variable picks the first nonconnected node by searching for an element value in the n matrix that is greater than zero and assigning j to that value.

The statement labeled with the number 2 initializes the variable i to 1, in effect, commencing the search of the connection matrix at Row 1. Next, the variable x is assigned the element value of the x matrix for Column j and Row i. When an element in the node matrix is greater than zero, a branch to the statement labeled with the number 3 occurs. Otherwise, the value of the variable i is incremented by 1 and a branch back to the statement labeled with the number 4 occurs.

The FOR-NEXT loop labeled with the number 3 searches the connection matrix for the smallest value. The statement IF n(i) = 0 THEN x(i,j) = 9999 assigns the relatively high value of 9999 to the element in the connection matrix when the node matrix element that matches the connection matrix row is zero. When this occurs, it indicates that the node is already connected, thus, assigning the value 9999 precludes the selection of the element. Next, the statement IF x(i,j) <=x AND n(j)>0 tests the value of the element against the first value in the column and the value of the node element associated with the column. When x(i,y) <=x, the selected connection matrix element is less than the prior selected element. When n(j)>0, the node is not yet connected. Thus, when both conditions are true, the next three statements are executed.

The first statement following the previously discussed IF statement assigns x to the value contained in the x(i,j) element as that is now the lowest legal value in the column. The next two statements assign the index of the smallest value in the column based upon the row and column of the element to the variables indexi and indexj, respectively. After the FOR-NEXT loop is completed, the value of the smallest value element is compared to 9999. If the smallest value element equals 9999, all elements in the connection matrix have been operated upon and the program terminates. Otherwise, the WEIGHT variable is incremented by the value assigned to x, the latter representing the lowest link value in the column just searched.

After displaying the selected link by indicating the nodes it connects and their link weight, the program sets the node element associated with the column just processed to zero. This is done to preclude the column from being searched again, because it indicates that the node represented by the column is connected by a link. The last FOR-NEXT loop, which uses i as an index, tests the node matrix for an element whose value is greater than zero because this condition indicates that one or more nodes remain to be connected. When this occurs, a branch to the statement labeled with the number 2 also occurs. Then, because $X_{ij} = X_{ji}$, X_{ji} is set to a value of 9999 to preclude its use, using the value of the variables indexj and indexi to represent the row and column, respectively. Once the column has been searched and the lowest value element selected, the row position will be used as a pointer for the next column to be searched. This is accomplished by the statement j = indexi, after which a branch back to the statement with the label number 1 occurs.

The execution of the modified MST program results in the selection of links between nodes based upon their link weight. Table 6.5 illustrates the execution of the program listed in Table 6.4. Note that the MST selected has a total link weight of 10, which is the lowest weight associated with a tree that connects all nodes and whose structure precludes a fan-out situation.

Table 6.5 MST Program Output

```
LINK SELECTION SUMMARY
LINK CONNECTS NODES 1   TO   2   LINK WEIGHT= 1
LINK CONNECTS NODES 2   TO   4   LINK WEIGHT= 2
LINK CONNECTS NODES 4   TO   5   LINK WEIGHT= 2
LINK CONNECTS NODES 5   TO   3   LINK WEIGHT= 5
TOTAL LINK WEIGHT =          10
```

6.3 CONSIDERING NETWORK CONSTRAINTS

The MST algorithm, although economically accurate, does not consider two key network constraints that could make its implementation impractical — the terminal response time of the locations interconnected and the capacity of a front-end processor or central server to service the total number of locations connected on one multidrop line.

6.3.1 Terminal Response Time Factors

Normally, full-screen display terminals are used on a multidrop circuit. The terminal response time is defined as the time from the operator pressing the Enter key to the first character appearing on the terminal's screen in response to the data sent to the computer. This response time depends upon a large set of factors, of which the major ones are listed in Table 6.6.

The line speed refers to the transmission data rate that determines how fast data can be transported between the terminal and the computer once the terminal is polled or selected. The type of transmission line, full- or half-duplex, determines whether or not an extra delay will be incurred to turn the line around after each poll. If a half-duplex transmission protocol is used, then the modem turnaround time will affect the terminal response time.

The number of characters serviced per poll refers to how the communications software services a terminal capable of storing 1920 characters on a full screen of 25 lines by 80 characters per line. To prevent one terminal from hogging the line, most communications software divides the screen into segments and services a portion of the screen during each poll sequence.

The type of polling can occur "round robin," where each terminal receives servicing in a defined order, or it can occur based upon a predefined priority. Although the computer processing time can greatly affect the terminal response time, it is normally beyond the control of the communications staff. The polling service time is the time it takes to poll

Table 6.6   Terminal Response Time Factors

Line speed
Type of transmission line
Modem turnaround time
Number of characters serviced per polling
Computer processing time
Polling service time

a terminal, so the communications software can service another segment of the screen when the data to be read or written to the terminal exceeds one segment. Finally, the probability of a transmission error occurring will affect the probability of transmitting the same data again, because detected errors are corrected by the retransmission of data.

6.3.2 Estimating Response Time

To estimate the average terminal response time requires an estimate of the average number of users that are using the terminals on a multidrop circuit. Next, the average number of characters to be transmitted in response to each depression of the Enter key must be estimated. This data can then be used to estimate the average terminal response time.

Suppose there are 10 terminals on a multidrop circuit and at any one time 4 are active, with approximately 10 lines of 30 characters on the display when the Enter key is pressed. If the communications software services segments of 240 characters, two polls will be required to service each terminal. Assuming a transmission rate of 4.8 kbps, which is equivalent to approximately 600 cps, it requires a minimum of 240/600 or 0.4 s to service the first segment on each terminal, excluding communications protocol overhead. Using a 25 percent overhead factor, which is normally reasonable for most protocols, the time to service the first segment becomes 0.5 s, resulting in the last terminal having its first segment serviced at a time of 2.0 s if all 4 active users requested servicing at the same time. Because 60 characters remain on each screen, the second poll requires 60/600 or 0.1 s per terminal plus 25 percent for overhead, or a total of 0.5 s until the fourth terminal is again serviced. Adding the time required to service each segment results in a total time of 2.5 s transpiring in the completion of the data transfer from the fourth terminal in the computer.

Now, assume that the average response is 300 characters. Then, the transmission of two screen segments is also required in the opposite direction. The first segment would then require 2.0 s for displaying on each terminal and the second segment requires 0.3 s plus 25 percent overhead or 0.375 s until the first character starts to appear on the fourth terminal, for a total response time (inbound and outbound) of 2.5 plus 2.375 or 4.875 s.

If a round robin polling sequence is used, the computer has an equal probability of polling any of the four terminals when the Enter key is pressed. Thus, the computed 4.875-s response time is the worst-case response time. The best-case response time would be the response time required to service the first terminal sending data, which in the previous example would be 2.1 s inbound and 2.0 s outbound until the first character is received or a total of 4.1 s. Thus, the average terminal response time would be (4.1 + 4.875)/2 or approximately 4.5 s.

6.3.3 Front-End Processing Limitations

A second limitation concerning the use of multidrop circuits is the capability of the front-end processor. In a large network that contains numerous terminal locations, the polling addressing capability of the front-end processor will limit the number of drops that can be serviced. Even if the processor could handle an infinite number of drops, polling delay times, as well as the effect of a line segment impairment breaking access to many drops, usually precludes most circuits to a maximum of 16 or 32 drops.

6.4 SUMMARY

When the number of drops in a network requires the use of multiple multidrop circuits, the network designer will normally consider the use of a more complex algorithm by the addition of a variety of constraints to the previously covered algorithms. Although such algorithms are best applied to network design problems by the use of computer programs, you can consider a practical alternative to these complex algorithms. This alternative is the subdivision of the network's terminal locations into servicing areas, based upon defining a servicing area to include a number of terminals that will permit an average response time that is acceptable to the end user. Then each segment can be analyzed using the MST algorithm to develop a minimum cost multidrop line to service all terminals within the servicing area.

7

SIZING COMMUNICATIONS EQUIPMENT AND LINE FACILITIES

Of the many problems associated with the acquisition of data communications networking devices, including LAN access controllers, multiplexers, and concentrators, one item often requiring resolution is the configuration or sizing of the device. The process of ensuring that the configuration of the selected device will provide a desired level of service is the foundation upon which the availability level of a network is built and, in many instances, is directly related to the number of dial-in lines connected to the device.

The appropriate sizing of LAN access controllers is an important consideration for a large number of organizations. Because LAN access controllers can be used by ISPs as a mechanism to provide dial access to the Internet, they represent one of the most commonly employed data communications devices used by ISPs. Because they also enable government agencies, academia, and private organizations to provide dial-in access to their LANs, the LAN access controller also represents a popular communications product used by nonservice providers. Although hardware manufacturers use the phrase "LAN access controller" to denote a specialized product for enabling dial network users to access a LAN, another equivalent device that requires sizing is a remote access server (RAS). Concerning the latter, RAS support is included in the popular Windows® NT® and Windows® 2000 operating systems. If you install RAS on a Windows NT or Windows 2000 server to enable employees or customers to access your server or LAN, you can use the information in this chapter to determine the number of dial-in modems and access lines to connect to your server. In a WAN environment, a similar sizing problem occurs when you need to provision a multiplexer, router, or switch to receive calls via the PSTN.

The failure to provide a level of access acceptable to network users can result in a multitude of problems. First, a user encountering a busy signal might become discouraged, take a break, or do something other than redial the telephone number of a network access port. Such action obviously will result in a loss of user productivity. If network usage is in response to customer inquiries, a failure to certify a customer purchase, return, reservation, or other action in a timely manner could result in the loss of customers to a competitor. This is similar to the situation where a long queue in front of a bank teller can result in the loss of customer accounts if the unacceptable level of service persists.

In this chapter, we will focus our attention upon the application of telephone traffic formulas to the sizing of data communications equipment and line facilities. Although most telephone traffic formulas were developed during the 1920s, many are applicable to such common problems as determining the number of dial-in business and WATS lines required to service remote PC users, as well as the number of ports or channels that should be installed in communications equipment connected to the dial-in lines. To appreciate the sizing process, we will first examine several methods that can be used to size equipment and line facilities. This will be followed by a detailed examination of the application of telephone traffic sizing formulas to data communications. More formally referred to as traffic dimensioning formulas, in this chapter we will examine the application of the Erlang B, Erlang C, and Poisson formulas to data communications equipment and facility sizing problems.

7.1 SIZING METHODS

There are many devices and line facilities that can be employed in a data communications network whose configuration or sizing problems are similar. Examples of line facilities include the number of dial-in local business and WATS lines required to be connected to telephone company rotaries, and examples of communications equipment sizing include determining the number of channels on LAN access controllers, multiplexers, data concentrators, and port selectors.

7.1.1 Experimental Modeling

Basically, two methods can be used to configure the size of communications network devices. The first method, commonly known as experimental modeling, involves the selection of the device configuration based

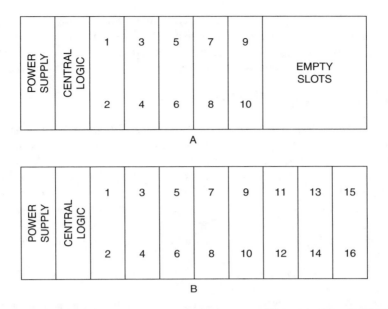

Figure 7.1   Experimental Modeling
Experimental modeling results in the adjustment of a network configuration based upon previous experience and gut intuition. (A) Initial configuration. (B) Adjusted configuration.

upon a mixture of previous experience and intuition. Normally, the configuration selected is less than the base capacity plus expansion capacity of the device. This enables the size of the device to be adjusted or upgraded without a major equipment modification if the initial sizing proved inaccurate. An example of experimental modeling is shown in Figure 7.1.

A rack-mounted LAN access controller is shown in Figure 7.1A. Initially, the controller was obtained with 5 dual-port adapters to support 10 ports of simultaneous operation. Assuming the base unit can support 8 dual-port adapters, if the network manager's previous experience or gut intuition proves wrong, the controller can be upgraded easily. This is shown in Figure 7.1B, where the addition of 3 dual-port adapters permits the controller to support 16 ports in its adjusted configuration.

Assuming each controller port is connected to a modem and business line, experimental modeling, although often representing a practical solution to equipment sizing, can also be expensive. If your organization began with the configuration shown in Figure 7.1B and adjusted the number of controller ports downward to that shown in Figure 7.1A, you would incur some cost for the extra modems and business lines, even if the manufacturer of the LAN access controller was willing to take back the port adapters you did not actually need. Thus, although experimental modeling is better than simply guessing, it can also result in the expenditure of funds for unnecessary hardware and communications facilities.

7.1.2 The Scientific Approach

The second method that can be employed to size network components ignores experience and intuition. This method is based upon knowledge of data traffic and the scientific application of mathematical formulas to traffic data. Hence, it is known as the scientific approach or method of equipment sizing. Although some of the mathematics involved in determining equipment sizing can become quite complex, a series of tables generated by the development of appropriate computer programs can be employed to reduce many sizing problems to one of a single table lookup process.

Although there are advantages and disadvantages to each method, the application of a scientific methodology to equipment sizing is a rigorously defined approach. Thus, there should be a much higher degree of confidence and accuracy of the configuration selected when this method is used. On the negative side, the use of a scientific method requires a firm knowledge or accurate estimate of the data traffic. Unfortunately, for some organizations, this may be difficult to obtain. In many cases, a combination of two techniques will provide an optimum situation. For such situations, sizing can be conducted using the scientific method with the understanding that the configuration selected may require adjustment under the experimental modeling concept. In the remainder of this chapter, we will focus our attention upon the application of the scientific methodology to equipment sizing problems.

7.2 TELEPHONE TERMINOLOGY RELATIONSHIPS

Most of the mathematics used for sizing data communications equipment evolved out of work originally performed to solve the sizing problems of telephone networks. From a discussion of a few basic telephone network terms and concepts, we will see the similarities between the sizing problems associated with data communications equipment and facilities and the structure of the telephone network. Building upon this foundation, we will learn how to apply the mathematical formulas developed for telephone network sizing to data communications network configurations.

To study the relationship between telephone network communications component sizing problems, let us examine a portion of the telephone network and study the structure and calling problems of a small segment formed by 2 cities, each assumed to contain 1000 telephone subscribers.

7.2.1 Telephone Network Structure

The standard method of providing an interconnection between subscribers in a local area is to connect each subscriber's telephone to what is known as the local telephone company exchange. Other synonymous terms for the local telephone company exchange include the local exchange and telephone company central office. When one subscriber dials another connected to the same exchange, the subscriber's call is switched to the called party number through the switching facilities of the local exchange. If we assume each city has one local exchange, then all calls originating in that city and to a destination located within that city will be routed through one common exchange.

Because our network segment selected for analysis consists of two cities, we will have two telephone company exchanges, one located in each city. To provide a path between cities for intercity calling, a number of lines must be installed to link the exchanges in each city. The exchange in each city can act then as a switch, routing the local subscribers in each city to parties in the other city.

7.2.1.1 Trunks and Dimensioning

As shown in the top part of Figure 7.2, a majority of telephone traffic in the network segment consisting of the two cities will be among the subscribers of each city. Although there will be telephone traffic between the subscribers in each city, it normally will be considerably less than the amount of local traffic in each city. The path between the two cities connecting their telephone offices is known as a trunk.

One of the many problems in designing the telephone network is determining how many trunks should be installed between telephone company exchanges. A similar sizing problem occurs many times in each city at locations where private organizations desire to install switchboards. An example of a sizing problem with this type of equipment is illustrated in the lower portion of Figure 7.2. In effect, the switchboard functions as a small telephone exchange, routing calls carried over a number of trunks installed between the switchboard and the telephone company exchange to a larger number of subscriber lines connected to the switchboard. The determination of the number of trunks required to be installed between the telephone exchange and the switchboard is called dimensioning and is critical for the efficient operation of the facility. If insufficient trunks are available, company personnel will encounter an unacceptable number of busy signals when trying to place an outside telephone call. Once again, this will obviously affect productivity.

Returning to the intercity calling problem, consider some of the problems that can occur in dimensioning the number of trunks between central

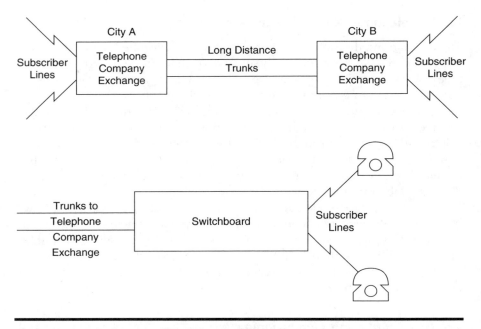

Figure 7.2 Telephone Traffic Sizing Problems
Although most subscriber calls are routed locally through the local telephone company exchange or local switchboard to parties in the immediate area, some calls require access to trunks. The determination of the number of trunks required to provide an acceptable grade of service is known as line dimensioning and is critical for the effective operation of the facility.

offices located in the two cities. Assume that based upon a previously conducted study, it was determined that no more than 50 people would want to have simultaneous telephone conversations where the calling party was in one city and the called party in the other city. If 50 trunks were installed between cities and the number of intercity callers never exceeded 50, at any moment the probability of a subscriber completing a call to the distant city would always be unity, always guaranteeing success. Although the service cost of providing 50 trunks is obviously greater than providing a lesser number of trunks, no subscriber would encounter a busy signal.

Because some subscribers might postpone or choose not to place a long-distance call at a later time if a busy signal is encountered, a maximum level of service will produce a minimum level of lost revenue. If more than 50 subscribers tried to simultaneously call parties in the opposite city, some callers would encounter busy signals once all 50 trunks were in use. Under such circumstances, the level of service would be such that not all subscribers are guaranteed access to the long-distance trunks and the probability of making a long-distance call would be less than unity. Likewise, because the level of service is less than that required to provide all callers with access to the long-distance trunks, the service cost is less than the service cost associated with providing users with a probability of unity in accessing trunks. Similarly, as the probability of successfully accessing the long-distance trunk decreases, the amount of lost revenue or customer waiting costs will increase. Based upon the preceding, a decision model factoring into consideration the level of service versus expected cost can be constructed as shown in Figure 7.3. The location where the total cost is minimal represents the optimum level of service one should provide.

7.2.1.2 The Decision Model

For the decision model illustrated in Figure 7.3, suppose the optimum number of trunks required to link the two cities is 40. The subscriber line-to-trunk ratio for this case would be 1000 lines to 40 trunks, for a 25:1 ratio.

To correctly dimension the optimum number of trunks linking the two cities requires an understanding both of economics and subscriber traffic. In dimensioning the number of trunks, a certain trade-off will result that relates the number of trunks or level of service to the cost of providing that service and the revenue lost by not having enough trunks to satisfy the condition when a maximum number of subscribers in one city dial subscribers in another. To determine the appropriate level of service, a

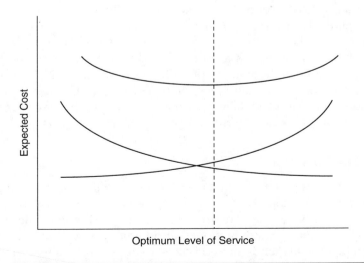

Figure 7.3 Using a Decision Model to Determine the Optimum Level of Service

decision model as illustrated in Figure 7.3 is required. Here, the probability of a subscriber successfully accessing a trunk corresponds to the level of service provided. As more trunks are added, the probability of access increases, as well as the cost of providing such access. Correspondingly, the waiting cost of the subscriber or the revenue loss to the telephone company decreases as the level of service increases, where the total cost represents the combination of service cost and waiting cost. The point where the cost is minimal represents the optimal number of trunks or level of service that should be provided to link the two cities.

From a LAN access perspective, a similar decision model can be constructed. However, instead of focusing upon accessing trunks, our concern would be oriented toward providing access to a LAN via the PSTN. If the number of ports, modems, and business lines equals the number of employees or subscribers of the organization, nobody would experience a busy signal; however, the cost of providing this level of capacity would be high and during a portion of the day, most of its capacity would more than likely be unused. As we reduce the number of ports, modems, and dial-in lines, the level of service decreases and eventually employee or subscriber waiting time results in either lost productivity or lost revenue. Thus, from a LAN access perspective, you would also seek to determine an optimum level of service.

7.3 TRAFFIC MEASUREMENTS

Telephone activity can be defined by the calling rate and the holding time, which is the duration of the call. The calling rate is the number of times a particular route or path is used per unit time period, and the holding time is the duration of the call on the route or path. Two other terms that warrant attention are the offered traffic and the carried traffic. The offered traffic is the volume of traffic routed to a particular telephone exchange during a predetermined time period, and the carried traffic is the volume of traffic actually transmitted through the exchange to its destination during a predetermined period of time.

7.3.1 The Busy Hour

The key factor required to dimension a traffic path is knowledge of the traffic intensity during the time period known as the busy hour (BH). Although traffic varies by day and time of day and is generally random, it follows a certain consistency one can identify. In general, traffic peaks prior to lunch time and then rebuilds to a second daily peak in the afternoon. The busiest one-hour period of the day is the BH. It is the BH traffic level that is employed in dimensioning telephone exchanges and transmission routes, because one wants to size the exchange or route with respect to its busiest period.

It is important to note that the BH can vary considerably between organizations. For example, an ISP might experience its heaviest traffic between 6 p.m. and 8 p.m., once subscribers return home from work, digest their supper, and then attempt to go online. In comparison, a government agency would more than likely have its BH occur during the day.

Telephone traffic can be defined as the product of the calling rate per hour and the average holding time per call. This measurement can be expressed mathematically as:

$$T = C \times D$$

Where:
C = calling rate per hour
D = average duration per call

Using the above formula, traffic can be expressed in call-minutes (CM) or call-hours (CH), where a CH is the quantity represented by one or more calls having an aggregate duration of one hour.

If the calling rate during the BH of a particular day is 500 and the average duration of each call is 10 minutes, the traffic flow or intensity would be 500 × 10, or 5000 CM, which would be equivalent to 5000/60, or approximately 83.3 CH.

7.3.2 Erlangs and Call-Seconds

The preferred unit of measurement in telephone traffic analysis is the erlang, named after A.K. Erlang, a Danish mathematician. The erlang is a dimensionless unit in comparison to the previously discussed CMs and CHs. It represents the occupancy of a circuit where one erlang of traffic intensity on one traffic circuit represents a continuous occupancy of that circuit.

A second term often used to represent traffic intensity is the call-second (CS). The quantity represented by 100 CSs is known as 1 CCS. Here, the first C represents the quantity 100 and comes from the French term *cent*. Assuming a one-hour unit interval, the previously discussed terms can be related to the erlang as follows:

$$1 \text{ erlang} = 60 \text{ call-minutes} = 36 \text{ CCS} = 3600 \text{ CS}$$

If a group of 20 trunks were measured and a call intensity of 10 erlangs determined over the group, then we would expect one half of all trunks

Table 7.1 Traffic Conversion Table

Dimension		Erlangs (intensity)	
Minutes	Hours	CHs (quantity)	CCS (quantity)
12	.2	0.2	6
24	.4	0.4	12
36	.6	0.6	18
48	.8	0.8	24
60	1.0	1.0	36
120	2.0	2.0	72
180	3.0	3.0	108
240	1.0	4.0	144
300	5.0	5.0	180
360	6.0	6.0	210
420	7.0	7.0	252
480	8.0	8.0	288
540	9.0	9.0	324
600	10.0	10.0	360
900	15.0	15.0	540
1200	20.0	20.0	720
1500	25.0	25.0	900
1800	30.0	30.0	1080
2100	35.0	35.0	1260
2400	40.0	40.0	1440
2700	45.0	45.0	1620
3000	50.0	50.0	1800
6000	100.0	100.0	3600

to be busy at the time of the measurement. Similarly, a traffic intensity of 600 CM or 360 CCS offered to the 20 trunks would warrant the same conclusion. Table 7.1 is a traffic conversion table that will facilitate the conversion of erlangs to CCS and vice versa. Because the use of many dimensioning tables is based upon traffic intensity in erlangs or CCS, the conversion of such terms frequently is required in the process of sizing facilities.

To illustrate the applicability of traffic measurements to a typical communications network configuration, assume your organization has a 10-position rotary connected to a LAN access controller. Further assume that you measured the number of calls and holding time per call during a one-hour period and determined the traffic distribution to be that illustrated in Figure 7.4. Note that the total holding time is 266 minutes,

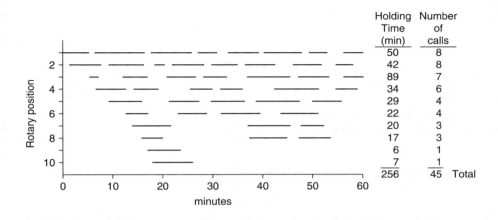

Holding Time (min)	Number of calls
50	8
42	8
89	7
34	6
29	4
22	4
20	3
17	3
6	1
7	1
256	45 Total

Figure 7.4   Traffic Distribution Example

or 4.43 hours. Thus, the average traffic intensity during this one-hour period is 4.43/1 or 4.43 erlangs.

During the BH illustrated in Figure 7.4, a total of 45 calls resulted in a cumulative holding time of 266 minutes. Thus, the average holding time per call is 266/45, or 5.91 minutes, which is equivalent to 0.0985 hours. Multiplying the average holding time (0.0985 hours) per call by the number of calls (45) results in the average traffic intensity of 4.43 erlangs. Note that this tells us that if we know the average holding time and the number of calls, we can easily determine the traffic intensity. The previously noted relationship between the average traffic intensity (E) and the holding times on each rotary position (H_i) during time period (T) can be expressed mathematically as follows:

$$E = \frac{\sum_{i=1}^{n} H_i}{T}$$

Substituting the data contained in Figure 7.4 we obtain for the one-hour period:

$$E = \frac{266}{1} = \text{call-minutes, or } 4.43 \text{ CH}$$

The average call holding time or average call duration (D) can be expressed in terms of the total holding time (3 H_i) and the number of calls (c) as:

$$D = \frac{\sum\limits_{i=1}^{n} H_i}{C} = \frac{266}{45} = 5.91$$

Because $\sum H_i = C \times D$ and as we previously noted, $E = \sum H_i / T$, we can express the traffic intensity (E) in terms of the number of calls, average call duration, and time period. Doing so we obtain:

$$E = \frac{C \times D}{T}$$

Using the data contained in Figure 7.4, we can compute the traffic intensity in erlangs as follows using a call duration expressed in minutes:

$$E = \frac{45 \times 5.91}{60} = 4.43 \text{ erlangs}$$

To find the traffic intensity when call duration is expressed in hours, our substitution would become:

$$E = \frac{45 \times 5.91/60}{1} = 4.43 \text{ erlangs}$$

As indicated, knowledge of the average call duration and number of calls permits the computation of traffic intensity. Because many types of network management systems as well as access controllers, multiplexers, and computer systems collect statistics to include the number of calls and call duration, it is often possible to obtain traffic intensity information. Even when you cannot obtain such information directly, it is often possible to obtain information indirectly. As an example, consider a 10-position rotary connected to modems that in turn are connected to ports on a LAN access controller. If the telephone company cannot provide the required information and your organization lacks monitoring equipment, the statistics you require may be obtainable from the access controller.

7.3.3 Grade of Service

One important concept in the dimensioning process is what is known as the grade of service. To understand this concept, let us return to our intercity calling example illustrated in Figure 7.2, again assuming 50 trunks are used to connect the telephone exchanges in each city. If a subscriber attempts to originate a call from one city to the other when all trunks are in use, that call is said to be blocked. Based upon mathematical formulas, the probability of a call being blocked can be computed given the traffic intensity and number of available trunks. The concept of determining the probability of blockage can be adapted easily to the sizing of data communications equipment.

From a logical analysis of traffic intensity, it follows that if a call were to be blocked, such blockage would occur during the BH, because that is the period when the largest amount of activity occurs. Thus, telephone exchange capacity is engineered to service a portion of the BH traffic, the exact amount of service depending upon economics, as well as the political process of determining the level of service one desires to provide to customers.

You could overdimension the route between cities and provide a trunk for every subscriber. This would ensure that a lost call could never occur and would be equivalent to providing a dial-in line for every terminal in a network. Because a 1:1 subscriber-to-trunk ratio is not economical and will result in most trunks being idle a large portion of the day, we can expect a lesser number of trunks between cities than subscribers. As the number of trunks decreases and the subscriber-to-trunk ratio correspondingly increases, we can intuitively expect some sizings to result in some call blockage. We can specify the number of calls we are willing to have blocked during the BH. This specification is known as the grade of service and represents the probability (P) of having a call blocked. If we specify a grade of service of 0.05 between the cities, we require a sufficient number of trunks so that only 1 call in every 20, or 5 calls in every 100, will be blocked during the BH.

7.3.4 Route Dimensioning Parameters

To determine the number of trunks required to service a particular route you can consider the use of several formulas. Each formula's use depends upon the call arrival and holding time distribution, the number of traffic sources, and the handling of lost or blocked calls. Regardless of the formula employed, the resulting computation will provide you with the probability of call blockage or grade of service based upon a given number of trunks and level of traffic intensity.

Concerning the number of traffic sources, you can consider the calling population as infinite or finite. If calls occur from a large subscriber population and subscribers tend to redial if blockage is encountered, the calling population can be considered as infinite. The consideration of an infinite traffic source results in the probability of a call arrival becoming constant and does not make the call dependent upon the state of traffic in the system. The two most commonly employed traffic-dimensioning equations are both based upon an infinite calling population.

Concerning the handling of lost calls, such calls can be considered cleared, delayed, or held. When calls are considered held, it is assumed that the telephone subscriber, upon encountering a busy signal, immediately redials the desired party. The lost call delayed concept assumes each subscriber is placed in a waiting mechanism for service and forms the basis for queuing analysis. Because we can assume a service or nonservice condition, we can disregard the lost call delayed concept unless access to a network resource occurs through a data private branch exchange (PBX) or port selector that has queuing capability.

7.3.5 Traffic Dimensioning Formulas

The principal traffic dimensioning formula used in North America is based upon the lost call concept and is commonly known as the Poisson formula. In Europe, traffic formulas are based upon the assumption that a subscriber encountering a busy signal will hang up the telephone and wait a certain amount of time prior to redialing. The Erlang B formula is based upon this lost call cleared concept.

7.4 THE ERLANG TRAFFIC FORMULA

The most commonly used telephone traffic dimensioning equation is the Erlang B formula. This formula is predominantly used outside the North American continent. In addition to assuming that data traffic originates from an infinite number of sources, this formula is based upon the lost call cleared concept. This assumption is equivalent to stating that traffic offered to but not carried by one or more trunks vanishes, and this is the key difference between this formula and the Poisson formula. The latter formula assumes that lost calls are held and it is used for telephone dimensioning mainly in North America. Because data communications system users can be characterized by either the lost call cleared or lost call held concept, both traffic formulas and their application to data networks will be covered in this chapter.

If E is used to denote the traffic intensity in erlangs and T represents the number of trunks, channels, or ports designed to support the traffic, the probability P(T,E) represents the probability that T trunks are busy when a traffic intensity of E erlangs is offered to those trunks. The probability is equivalent to specifying a grade of service and can be expressed by the Erlang traffic formula as follows:

$$P(T,E) = \frac{\dfrac{E^T}{T!}}{1 + \dfrac{E^1}{1!} + \dfrac{E^2}{2!} + \dfrac{E^3}{3!} + \ldots \dfrac{E^T}{T!}} = \frac{\dfrac{E^T}{T!}}{\displaystyle\sum_{t=0}^{T} \dfrac{E^i}{i!}}$$

Where:

$$T! = T \times (T-1) \times (T-2) \ldots 3 \times 2 \times 1$$

And

$$0! = 1$$

Table 7.2 presents a list of factorials and their values to assist you with computing specific grades of service based upon a given traffic intensity and trunk quantity.

Table 7.2 Factorial Values

N	Factorial N	N	Factorial N
1	1	51	1.551118753287382E+66
2	2	52	8.065817517094390E+67
3	6	53	4.274883284060024E+69
4	24	54	2.308436973392413E+71
5	120	55	1.269640335365826E+73
6	720	56	7.109985878048632E+74
7	5040	57	4.052691950487723E+76
8	40320	58	2.350561331282879E+78
9	362880	59	1.386831185456898E+80
10	3628800	60	8.320987112741390E+81
11	39916800	61	5.075802138772246E+83
12	479001600	62	3.146997326038794E+85
13	6227020800	63	1.982608315404440E+87
14	87178291200	64	1.268869321858841E+89
15	1307674368000	65	8.247650592082472E+90
16	20922789888000	66	5.443449390774432E+92
17	355687428096000	67	3.647111091818871E+94
18	6402373705728000	68	2.480035542436830E+96
19	1.216451004088320E+17	69	1.711224524281413E+98
20	2.432902008176640E+18	70	1.197857166966989E+100
21	5.109094217170944E+19	71	8.504785885678624E+101

Table 7.2   Factorial Values (continued)

N	Factorial N	N	Factorial N
22	1.124000727777608E+21	72	6.123445837688612E+103
23	2.585201673888498E+22	73	4.470115461512686E+105
24	6.204484017332394E+23	74	3.307885441519387E+107
25	1.551121004333098E+25	75	2.480914081139540E+109
26	4.032914611266057E+26	76	1.855494701666051E+111
27	1.088886945041835E+28	77	1.451830920282859E+113
28	3.048883446117138E+29	78	1.132428117820629E+115
29	8.841761993739701E+30	79	8.946182130782980E+116
30	2.652528598121911E+32	80	7.156945704626380E+118
31	8.222838654177924E+33	81	5.797126020747369E+120
32	2.631308369336936E+35	82	4.753643337012843E+122
33	8.683317618811889E+36	83	3.045523969720660E+124
34	2.952327990396041E+38	84	3.314240134565354E+126
35	1.033314796638614E+40	85	2.817104114380549E+128
36	3.719933267899013E+41	86	2.422709538367274E+130
37	1.376375309122635E+43	87	2.107757298379527E+132
38	5.230226174666010E+44	88	1.854826422573984E+134
39	2.036788208119745E+46	89	1.650795516090847E+136
40	8.159152832478980E+47	90	1.485715964481761E+138
41	3.345252661316380E+49	91	1.352001527678403E+140
42	1.405006117752880E+51	92	1.243841405464131E+142
43	6.041526306837884E+52	93	1.156772507081641E+144
44	2.658271574788450E+54	94	1.087366156656743E+146
45	1.196222208654802E+56	95	1.032997848823906E+148
46	5.502622159812089E+57	96	9.916779348709491E+149
47	2.586232415111683E+59	97	9.619275968248216E+151
48	1.241391559253607E+61	98	9.426890448883248E+153
49	6.082818640342679E+62	99	9.332621544394415E+155
50	3.041409320171338E+64	100	9.332621544394418E+157

To illustrate the use of the Erlang traffic formula, assume that a traffic intensity of 3 erlangs is offered to a three-position rotary. The grade of service is calculated as follows:

$$P(T,E) = \frac{\dfrac{E^T}{T!}}{\displaystyle\sum_{t-0}^{T} \dfrac{E^i}{T!}} = \frac{\dfrac{3^3}{3 \times 2 \times 1}}{\dfrac{3^0}{0!} + \dfrac{3^1}{1!} + \dfrac{3^2}{2!} + \dfrac{3^3}{3!}} = 0.346$$

This means that on the average during the BH, 34.6 out of every 100 calls will encounter a busy signal and for most organizations will represent an undesirable grade of service.

7.4.1 Computing Lost Traffic

Based upon the computed grade of service, we can compute the traffic lost during the BH. Here, the traffic lost (e) is the traffic intensity multiplied by the grade of service. Thus, the traffic lost by Position 3 is:

$$e_3 = E \times P(3,3) = 3 \times 0.345 = 1.038 \text{ erlangs}$$

Now, let us assume the rotary is expanded to four positions. The grade of service then becomes:

$$P(4,3) = \frac{\dfrac{3^4}{4 \times 3 \times 2 \times 1}}{\dfrac{3^0}{0!} + \dfrac{3^1}{1!} + \dfrac{3^2}{2!} + \dfrac{3^3}{3!} + \dfrac{3^4}{4!}} = 0.2061$$

This expansion improves the grade of service so that approximately one in five calls now receives a busy signal during the BH. The traffic lost by Position 4 now becomes:

$$e_4 = E \times P(4,3) = 3 \times 0.2061 = 0.6183 \text{ erlangs}$$

Note that the traffic carried by the fourth position is equal to the difference between the traffic lost by the three-position rotary and the traffic lost by the four-position rotary. That is:

Traffic carried by position 4 = 1.038 − 0.6183 = 0.4197 erlangs

Based upon the preceding, we can calculate both the traffic carried and the traffic lost by each position of an n-position rotary. The results of the traffic computations are obtainable once we know the number of positions on the rotary and the traffic intensity offered to the rotary group. As noted before, the traffic lost by position $n(e_n)$ can be expressed in terms of the grade of service and traffic intensity as follows:

$$e_n = E \times P(T_n, E)$$

Substituting the preceding in the Erlang formula gives:

$$e_n = E \times \frac{\dfrac{E^n}{n!}}{\displaystyle\sum_{t=0}^{n} \dfrac{E^i}{i!}}$$

where e_n is the traffic lost by the nth position on the rotary.

Because the traffic carried by any rotary position is the difference between the traffic offered to the position and the traffic lost by the position, we can easily compute the traffic carried by each rotary position. To do so, let us proceed as follows.

Let e_{n-1} equal the traffic lost by position n − 1.

Then, e_{n-1} becomes the traffic that is offered to position n on the rotary. Thus, the traffic carried by position n is equivalent to $e_{n-1} - e_n$. In the case of the first rotary position on a four-position rotary, the traffic lost becomes:

$$e_1 = E \times \frac{\dfrac{E^1}{1!}}{1 + \dfrac{E^1}{1!}} = \frac{E^2}{1 + E}$$

Then, the traffic carried by the first rotary position is the difference between the traffic intensity offered to the rotary group (E) and the traffic lost by the first position. That is, if T_{Cn} is the traffic carried by position n, then:

$$T_{C1} = E - e_1 = E - \frac{E^2}{1 + E} = \frac{E}{1 + E}$$

For the second rotary position, traffic lost by that position is:

$$e_1 = E \times \frac{\dfrac{E^2}{2!}}{1 + \dfrac{E^1}{1!} + \dfrac{E^2}{2!}} = \frac{E^3}{2 + 2E + E^2}$$

Then, the traffic carried by the second position on the rotary is $e_1 - e_2$ or:

$$e_1 - e_2 = \frac{E^2}{1 + E} - \frac{E^2}{2 + 2E + E^3}$$

Table 7.3 Traffic Lost and Traffic Carried by Rotary Position

Rotary Position	Traffic Lost	Traffic Carried
1	$e_1 = \dfrac{E^2}{1+E}$	$E - e_1 = \dfrac{E}{1+E}$
2	$e_1 = \dfrac{E^3}{2^2 + E^2}$	$e_1 - e_2 = \dfrac{E^2}{1+E} + \dfrac{E^3}{2 + 2E + E^2}$
3	$e_3 = \dfrac{E^4}{6 + 6E + 3E^2 + E^3}$	$e_1 - e_3 = \dfrac{E^3}{2 + 2E + E^2} + \dfrac{E^4}{6 + 6E + 5E^2 + E^3}$
4	$e_1 = \dfrac{E^5}{24 + 24E + 12E^2 + 4E^3 + E^4}$	$e_3 - e_4 = \dfrac{E^4}{6 + 6E + 5E^2 + E^3}$ $+ \dfrac{E^5}{24 + 24E + 12E^2 + 4E^3 + E^4}$

We can continue this process to compute both the traffic carried and the traffic lost by each rotary position. Table 7.3 summarizes the formulas used to obtain the traffic lost and traffic carried for each position of a four-position rotary group.

7.4.2 Traffic Analysis Program

To assist with the computations required to determine the grade of service and traffic distribution over each port on a rotary group, a program was developed using the Microsoft QuickBasic compiler. Table 7.4 contains the listing of the traffic analysis program that can be used to analyze rotaries containing up to 60 positions. For rotaries beyond 60 positions, the program can be altered; however, execution time will considerably increase.

If you are not familiar with the QuickBasic compiler, several entries in the program listing contained in Table 7.4 may warrant an explanation. Due to this, we will examine the program listing to provide a firm understanding of statements that may be different from the basic interpreter or compiler they are using, as well as to obtain a better understanding of the logical construction of the program.

The $DYNAMIC statement in the second program line allocates memory to arrays as required. The FACTORIAL# statement allocates 61 elements (0 through 60) for the array that will contain the values of factorial 0

Table 7.4 Traffic Analyzer Program Listing

```
REM Traffic Analyzer Program
REM $DYNAMIC
DIM FACTORIAL#(60)
DIM TL#(60), TC#(60)
REM E is the offered load in Erlangs
REM PORT is the number of ports, dial in lines or trunks
REM GOS is the grade of service for port or channel PORT with
traffic E
REM TL# is an array that holds traffic lost by port number
REM TC# is an array that holds traffic carried by port number
CLS
PRINT TAB(25); "Traffic Analyzer"
PRINT
PRINT "This program computes the grade of service and the
traffic carried"
PRINT "and lost by each port or channel in an n position rotary
type group"
PRINT
INPUT "Enter traffic intensity in Erlangs"; E#
1 INPUT "Enter number of ports -maximum 60"; PORT
IF PORT > 60 OR PORT < 1 GOTO 1
GOSUB 100   'compute factorial 1 TO PORT
REM Compute the grade of service
                PORT# = PORT
                N# = (E# ^ PORT#) / FACTORIAL#(PORT)
                        D# = 0
                        FOR S = 0 TO PORT
                        S# = S
                        D# = D# + (E# ^ S#) / FACTORIAL#(S)
                        NEXT S
                GOS# = N# / D#
REM Compute the traffic lost by port
        FOR S = 1 TO PORT
                S# = S
                LN# = E# * (E# ^ S# / FACTORIAL#(S))
                LD# = 0
                        FOR S1 = 0 TO S
                        S1# = S1
                        LD# = LD# + E# ^ S1# / FACTORIAL#(S1)
                        NEXT S1
                        TL#(S) = LN# / LD#
                NEXT S
REM Compute the traffic carried by port
        FOR I = 1 TO PORT
                IF I = 1 THEN
                        TC#(I) = E# - TL#(1)
                        ELSE
                        TC#(I) = TL#(I - 1) - TL#(I)
                        END IF
        NEXT I
REM Output results
PRINT
```

Table 7.4   Traffic Analyzer Program Listing (continued)

```
PRINT "TOTAL TRAFFIC OFFERED"; E#; "ERLANGS TO"; PORT; "PORTS
PROVIDES A";
PRINT USING "##.####"; GOS#;
PRINT " GRADE OF SERVICE"
PRINT
PRINT TAB(25); "TRAFFIC DISTRIBUTION"
PRINT
PRINT "PORT#    TRAFFIC OFFERED    TRAFFIC CARRIED      TRAFFIC
LOST"
PRINT
FOR I = 1 TO PORT
        PRINT USING "##"; I;
        PRINT USING "           ###.#####"; E#;
        PRINT USING "           ###.#####"; TC#(I);
        PRINT USING "           ###.#####"; TL#(I)
        E# = E# - TC#(I)
NEXT I
PRINT
PRINT "TRAFFIC LOST BY LAST PORT IS ";
PRINT USING "###.##### "; TL#(PORT);
PRINT "ERLANGS"
PRINT
PRINT "GRADE OF SERVICE IS EQUIVALENT TO 1 IN ";
PRINT USING "#####  "; INT((1 / GOS#) + .5);
PRINT "CALLS RECEIVING A BUSY SIGNAL"
END
REM subroutine to compute factorials
100 FOR I = 1 TO PORT
        p# = 1
                FOR J = I TO 1 STEP -1
                p# = p# * J
                NEXT J
        FACTORIAL#(I) = p#
        NEXT I
        FACTORIAL#(0) = 1
        RETURN
```

through factorial 60. Note that the variable suffix # (hash sign) is used in Basic to denote a double precision variable. Similar to FACTORIAL#, TL#, and TC# are arrays that are used to hold the double precision values of traffic lost and traffic carried by each port.

After the traffic intensity in erlangs (assigned to the variable E#) and the number of ports (assigned to the variable PORT) are entered, the program branches to the subroutine beginning at Statement Number 100. This subroutine computes the values of Factorial 0 through the number assigned to PORT and stores those factorial values in the array FACTO-RIAL#.

Table 7.5   Traffic Analyzer Program Execution

```
                        Traffic Analyzer


This program computes the grade of service and the traffic
carried and lost by each port or channel in an n position
rotary type group

Enter traffic intensity in Erlangs? 3
Enter number of ports -maximum 60? 4

TOTAL TRAFFIC OFFERED 3 ERLANGS TO 4 PORTS PROVIDES A 0.2061
GRADE OF SERVICE

                    TRAFFIC DISTRIBUTION

PORT#    TRAFFIC OFFERED    TRAFFIC CARRIED     TRAFFIC LOST

  1          3.00000           0.75000           2.25000
  2          2.25000           0.66176           1.58824
  3          1.58824           0.54977           1.03846
  4          1.03846           0.42014           0.61832

TRAFFIC LOST BY LAST PORT IS   0.61832 ERLANGS

GRADE OF SERVICE IS EQUIVALENT TO 1 IN   5   CALLS RECEIVING A
BUSY SIGNAL
```

After computing the factorial values, the program computes the grade of service using the equations previously described in this chapter. Similarly, the traffic lost and carried by each port is computed by computerizing the previously described equations to Basic language statements.

To illustrate the equivalency of a grade of service (stored in the variable GOS#) to 1 in N calls obtaining a busy signal, GOS# is first divided into 1. Next, 0.5 is added to the result to raise its value to the next highest number prior to taking the integer value of the computation. This is necessary because the INT function rounds down the result obtained by dividing GOS# into unity.

The result of the execution of the traffic analyzer program using a traffic intensity of 3 erlangs being presented to a four-position rotary is contained in Table 7.5. Note that the grade of service is 0.2061, which is approximately equivalent to one in five calls receiving a busy signal.

Through the use of the traffic analyzer program you can vary the traffic intensity or the number of ports on the rotary group to study the resulting traffic distribution and grade of service. To illustrate this, assume you want to analyze the effect of increasing the rotary group to five positions. Here, you could simply rerun the traffic analyzer program as illustrated in Table

Table 7.6   Analyzing the Effect of Port Expansion

```
                        Traffic Analyzer


This program computes the grade of service and the traffic
carried and lost by each port or channel in an n position
rotary type group

Enter traffic intensity in Erlangs? 3
Enter number of ports -maximum 60? 5

TOTAL TRAFFIC OFFERED 3 ERLANGS TO 5 PORTS PROVIDES A 0.1101
GRADE OF SERVICE

                      TRAFFIC DISTRIBUTION

PORT#    TRAFFIC OFFERED    TRAFFIC CARRIED    TRAFFIC LOST

 1           3.00000            0.75000           2.25000
 2           2.25000            0.66176           1.58824
 3           1.58824            0.54977           1.03846
 4           1.03846            0.42014           0.61832
 5           0.61832            0.28816           0.33016

TRAFFIC LOST BY LAST PORT IS    0.33016 ERLANGS

GRADE OF SERVICE IS EQUIVALENT TO 1 IN    9    CALLS RECEIVING A
BUSY SIGNAL
```

7.6. Note that when the rotary group is expanded to five positions, the grade of service is approximately equivalent to one in nine calls receiving a busy signal. In addition, you can use multiple executions of the traffic analyzer program to determine the change in the traffic lost by the last port in a port grouping as you increase or decrease the number of ports to service a given traffic intensity. In comparing the executions of the program displayed in Table 7.5 and Table 7.6, note that an increase in the number of ports from three to five decreased the traffic lost by the last port from 0.61832 to 0.33016 erlangs. Thus, you can use this program as a devil's advocate to determine what-if information without having to actually install or remove equipment and perform the line measurements normally associated with sizing such equipment.

7.4.3   Traffic Capacity Planning

There are three methods by which the erlang distribution equation can be used for capacity planning purposes. The first method, as previously illustrated, uses the erlang distribution equation to compute a grade of

Table 7.7 Traffic Capacity Planner Program

```
REM Traffic Capacity Planner Program
REM $DYNAMIC
DIM FACTORIAL#(60)
DIM E#(4, 40)
DIM GOS(10)
REM E is the offered load in Erlangs
REM E#(I,S) contains resulting traffic for GOS of I when S ports used
REM PORT is the number of ports, dial in lines or trunks
REM GOS is the grade of service for port or channel PORT with traffic E
MAXPORT = 40
CLS
FOR I = 1 TO 4
READ GOS(I)
NEXT I
DATA .01,.02,.04,.08
LPRINT TAB(25); "Capacity Planner"
LPRINT
LPRINT "This program computes and displays a table containing the traffic"
LPRINT "carrying capacity for a group of ports that will result in"
LPRINT "              a predefined grade of service"
LPRINT
        LPRINT
        LPRINT " NUMBER OF PORTS            TRAFFIC SUPPORTED IN ERLANGS PER
PORT"
        LPRINT "                           FOR INDICATED GRADE OF
SERVICE"
        LPRINT TAB(30);
        LPRINT USING "#.###   #.###   "; GOS(1); GOS(2);
        LPRINT USING "#.###   #.### "; GOS(3); GOS(4)
        LPRINT
GOSUB 100   'compute factorial 1 TO factorial MAXPORT
REM vary grade of service from .01 to .08 or 1 in 100 to 1 in 12.5 calls
busy
        FOR I = 1 TO 4

        GOS# = GOS(I)
        REM Vary ports from 1 to MAXPORT
            FOR PORT = 1 TO MAXPORT
                    REM Find traffic in Erlangs that provides GOS
                    LOW# = 1
                    HIGH# = 100000
                      TRY# = (LOW# + HIGH#) / 2
1  E# = TRY# / 1000
PORT# = PORT
                    N# = (E# ^ PORT#) / FACTORIAL#(PORT)
                    D# = 0
                    FOR S = 0 TO PORT
                    S# = S
                    D# = D# + (E# ^ S#) / FACTORIAL#(S)
                    NEXT S
                    IF ABS(GOS# - N# / D#) < .0005 THEN GOTO 5
                    IF GOS# - N# / D# < 0 THEN
                            OLD# = TRY#
                            TRY# = TRY# - ((TRY# - LOW#) / 2)
                            GOTO 1
                     ELSEIF GOS# - N# / D# > 0 THEN
                            TRY# = (TRY# + OLD#) / 2
                            GOTO 1
                    END IF
```

Table 7.7   Traffic Capacity Planner Program (continued)

```
5 E#(I, PORT) = TRY# / 1000
NEXT PORT
                    NEXT I
REM output results
      FOR S = 1 TO MAXPORT
              LPRINT TAB(4); S;
               LPRINT TAB(29);
               LPRINT USING "##.###  ##.###  "; E#(1, S); E#(2, S);
               LPRINT USING "##.###  ##.###"; E#(3, S); E#(4, S)
               NEXT S
      END
REM subroutine to compute factorials
100 FOR I = 1 TO MAXPORT
      p# = 1
              FOR J = I TO 1 STEP -1
              p# = p# * J
              NEXT J
      FACTORIAL#(I) = p#
      NEXT I
      FACTORIAL#(0) = 1
      RETURN
```

service based upon a defined traffic intensity and number of ports or channels. Using the value of the computed grade of service, you can then accept it or alter the traffic intensity or number of ports to obtain a desired grade of service.

The second method by which the erlang formula can be used is to determine the amount of traffic that can be serviced by a given number of ports or channels to provide a predefined grade of service. Using the erlang formula in this manner involves a trial and error process because different traffic intensity values must be substituted into the formula to determine if it results in the desired grade of service. Because this process can be quite laborious, a computer program was developed to generate a table of traffic intensities that can be serviced by a varying number of ports or channels to provide predefined grades of service.

Table 7.7 contains a program listing of a Traffic Capacity Planner Program, which was also developed using the QuickBasic Compiler. This program computes and displays the traffic intensity that can be offered to 1 to 40 ports to obtain 0.01, 0.02, 0.04, and 0.08 grades of service. You can easily vary both the grades of service or number of ports.

To vary the grades of service, the DATA statement should be changed. To increase the number of ports, the variable MAXPORT's value of 40 should be changed. When the number of grades of service or number of ports are increased, the DIM E#(4,40) statement should be increased to reflect the revised number of grades of service or ports for which the traffic intensity is to be computed. If the number of ports increases beyond

Table 7.8   Revised Traffic Capacity Planner Program

```
REM Traffic Capacity Planner Program
REM $DYNAMIC
DIM FACTORIAL#(60)
DIM E#(4, 40)
DIM GOS(4)
REM E is the offered load in Erlangs
REM E#(I,S) contains resulting traffic for GOS of I when S ports used
REM PORT is the number of ports, dial in lines or trunks
REM GOS is the grade of service for port or channel PORT with traffic E
MAXPORT = 44
CLS
FOR I = 1 TO 4
READ GOS(I)
NEXT I
DATA -.01,-.005,.00,.005
FOR TT = 1 TO 25
FOR K = 1 TO 4
GOS(K) = GOS(K) + 2 / 100
NEXT K
LPRINT TAB(30); "Capacity Planner"
LPRINT
LPRINT "THE FOLLOWING TABLE DISPLAYS THE TRAFFIC CARRYING CAPACITY FOR
THE"
LPRINT "INDICATED NUMBER OF PORTS THAT WILL RESULT IN A PREDEFINED GRADE
OF SERVICE"
LPRINT
        LPRINT
        LPRINT " NUMBER OF PORTS            TRAFFIC SUPPORTED IN ERLANGS PER
PORT"
        LPRINT "                               FOR INDICATED GRADE OF
SERVICE"
        LPRINT TAB(30);
        LPRINT USING "#.###   #.###   "; GOS(1); GOS(2);
        LPRINT USING "#.###   #.### "; GOS(3); GOS(4)
        LPRINT
GOSUB 100   'compute factorial 1 TO factorial MAXPORT
REM vary grade of service
        FOR I = 1 TO 4
        GOS# = GOS(I)
        REM Vary ports from 1 to MAXPORT
            FOR PORT = 1 TO MAXPORT
                        REM Find traffic in Erlangs that provides GOS
                        LOW# = 1
                        HIGH# = 100000
                          TRY# = HIGH#
1   E# = TRY# / 1000
PORT# = PORT
                        N# = (E# ^ PORT#) / FACTORIAL#(PORT)
                        D# = 0
                        FOR S = 0 TO PORT
                        S# = S
                        D# = D# + (E# ^ S#) / FACTORIAL#(S)
                        NEXT S
                        IF ABS(GOS# - N# / D#) < .0005 THEN GOTO 5
                        IF GOS# - (N# / D#) < 0 THEN
                                    HIGH# = TRY#
                                    TRY# = (TRY# - LOW#) / 2
                                    GOTO 1
                          ELSEIF GOS# - N# / D# > 0 THEN
                                LOW# = TRY#
                                TRY# = (TRY# + HIGH#) / 2
                                GOTO 1
```

Table 7.8   Revised Traffic Capacity Planner Program (continued)

```
                              END IF
5 E#(I, PORT) = TRY# / 1000
NEXT PORT
                        NEXT I
REM output results
        FOR S = 1 TO MAXPORT
                LPRINT TAB(4); S;
                LPRINT TAB(29);
                LPRINT USING "##.###  ##.###  "; E#(1, S); E#(2, S);
                LPRINT USING "##.###  ##.###"; E#(3, S); E#(4, S)
                NEXT S
        REM Skip to next page
        FOR I = 1 TO 12
        LPRINT
        NEXT I
NEXT TT
END
REM subroutine to compute factorials
100 FOR I = 1 TO MAXPORT
        p# = 1
                FOR J = I TO 1 STEP -1
                p# = p# * J
                NEXT J
        FACTORIAL#(I) = p#
        NEXT I
        FACTORIAL#(0) = 1
        RETURN
```

60, you should increase the size of the FACTORIAL#(60) array and be patient, as the computations become lengthy. A word of caution is in order if you require an expansion of the size of arrays. If the total number of elements in your program will exceed 64K, you must use the /AH option when invoking the QuickBasic compiler. Refer to Microsoft's Quick-Basic manual for information concerning the use of the /AH option.

The modifications required to change the program to compute the traffic supported by 1 to 44 ports or channels for grades of service ranging from 0.01 to 0.55 or 1 in 100 to 55 in 100 calls receiving a busy signal in increments of 0.005 are contained in Table 7.8. This illustration contains the revised Traffic Capacity Planner Program listing and can be compared to the program listing contained in Table 7.7 to denote the use of two additional FOR-NEXT statements that permit the use of only one DATA statement.

To speed up the computations of the trial and error procedure, the program was written to increment or decrement trials by one half of the previously used value. When the grade of service and the computed grade of service differ by less than 0.0005, a match is considered to have occurred and the traffic intensity used to compute the grade of service is placed into the E# array.

Table 7.9 illustrates the output produced from the execution of the Capacity Planner Program listed in Table 7.7. In examining the traffic

Table 7.9   Capacity Planner Program Execution

Capacity Planner

This program computes and displays a table containing the traffic
carrying capacity for a group of ports that will result in
a predefined grade of service

| NUMBER OF PORTS | TRAFFIC SUPPORTED IN ERLANGS PER PORT | | | |
| | FOR INDICATED GRADE OF SERVICE | | | |
	0.010	0.020	0.040	0.080
1	0.010	0.020	0.042	0.086
2	0.151	0.221	0.332	0.514
3	0.450	0.599	0.812	1.127
4	0.880	1.100	1.398	1.849
5	1.368	1.649	2.052	2.629
6	1.924	2.271	2.773	3.462
7	2.484	2.931	3.517	4.327
8	3.126	3.610	4.293	5.216
9	3.786	4.327	5.088	6.105
10	4.471	5.088	5.891	7.032
11	5.128	5.860	6.733	7.966
12	5.860	6.593	7.571	8.906
13	6.593	7.393	8.449	9.857
14	7.335	8.204	9.303	10.832
15	8.076	9.012	10.175	11.782
16	8.871	9.857	11.045	12.772
17	9.614	10.683	11.969	13.763
18	10.431	11.491	12.877	14.727
19	11.176	12.306	13.735	15.739
20	12.016	13.184	14.668	16.731
21	12.772	14.063	15.585	17.741
22	13.624	14.901	16.480	18.751
23	14.420	15.770	17.442	19.751
24	15.262	16.633	18.386	20.780
25	16.151	17.510	19.252	21.790
26	17.030	18.386	20.188	22.802
27	17.881	19.227	21.192	23.795
28	18.604	20.188	22.090	24.805
29	19.557	21.027	23.072	25.855
30	20.348	21.876	24.030	26.876
31	21.192	22.892	24.903	27.906
32	21.973	23.748	25.855	28.936
33	22.892	24.610	26.821	29.987
34	23.841	25.544	27.797	30.986
35	24.610	26.368	28.712	32.048
36	25.544	27.363	29.685	33.070
37	26.368	28.263	30.642	34.069
38	27.247	29.106	31.664	35.157
39	28.126	30.045	32.557	36.187
40	28.877	31.047	33.526	37.208

support by grade of service, you will note that a large group of ports is more efficient with respect to their traffic capacity support for a given grade of service than small groups of ports. Similarly, a small reduction in the number of ports from a large group of ports has a much more pronounced effect upon traffic capacity support than a similar reduction in the number of ports from a smaller group of ports. To illustrate the preceding, consider the 0.01 grade of service. Four groups of 10 ports support a total traffic intensity of 17.684 (4.471×4) erlangs. In comparison, one 40-port group supports a total of 28.877 erlangs. Based upon this, it is more than efficient to have one large rotary group than several smaller rotary groups. Consider this important concept of equipment sizing prior to breaking rotary or port groups into subgroups designed to service individual groups of end users. This concept also explains why it would be better to have one rotary group connected to V.90 modems operating at 56 kbps that can also service end-user 33.6 kbps transmission requirements than separate rotary groups.

7.4.4   Traffic Tables

A third method by which the erlang formula can be used is to generate a series of tables that indicate grades of service based upon specific traffic loads and a given number of ports or channels. Once again, a computer program was developed to facilitate the required computations.

Table 7.10 contains the QuickBasic program listing of a program that was written to compute a table of grades of service based upon a given traffic intensity and port or channel size using the erlang distribution.

The execution of the ERLANG.BAS program results in the generation of a data file named ERLANG.DAT that contains the grades of service for traffic intensities ranging from 0.5 to 40 erlangs for groups of up to 60 ports or channels. This program, like other programs developed to assist in traffic computations, can be easily modified to obtain grades of service for a different range of traffic intensities or larger number of ports or channels. The file ERLANG.DAT represents a table that lists grades of service based upon traffic intensities up to 77.5 erlangs in increments of 0.5 erlangs. Because the data file is in ASCII format, you can use any data processor capable of reading ASCII files to manipulate the entries to meet your specific requirements or you can use the Disk Operating System (DOS) redirect feature to print the file. Concerning the latter, you can enter the DOS command:

```
TYPE ERLANG.DAT > LPT1:
```

to direct the contents of the file ERLANG.DAT to your printer. These tables can be used to reduce many sizing problems to a simple lookup procedure

Table 7.10   Program to Generate Table of Grades of Service Using the Erlang Distribution

```
REM $DYNAMIC
DIM FACTORIAL#(60)
DIM E#(80), B#(80, 80)
REM E is the offered load in Erlangs
REM S is the number of ports, dial in lines or trunks
REM B(i,j) contains grade of service for port or channel i with
traffic j
OPEN "D:ERLANG.DAT" FOR OUTPUT AS #1
GOSUB 100   'compute factorial 1 TO 60
C = 0
FOR I = 5 TO 400 STEP 5
        C = C + 1
        E# = I / 10
E#(C) = E#
                FOR S = 1 TO 60
                    SX# = S
                    N# = (E# ^ SX#) / FACTORIAL#(S)
                    D# = 1
                    FOR D1 = 1 TO S
                    D1X# = D1
                    D# = D# + (E# ^ D1X#) / FACTORIAL#(D1)
                    NEXT D1
                B#(S, C) = N# / D#
                NEXT S
  NEXT I
  FOR I = 1 TO 80 STEP 5   'print 16 pages 5 entries per page
        PRINT #1, "                       ERLANG B DISTRIBUTION"
        PRINT #1, "        PROBABILITY ALL PORTS ARE BUSY WHEN
CALL ATTEMPTED"
        PRINT #1, "                       WHICH IS THE GRADE OF
SERVICE"
        PRINT #1,
        PRINT #1, "PORT #                 TRAFFIC IN ERLANGS"
        PRINT #1,
        PRINT #1, USING "      ##.##         ##.##"; E#(I);
E#(I + 1);
        PRINT #1, USING "      ##.## "; E#(I + 2);
        PRINT #1, USING "      ##.##         ##.##"; E#(I + 3);
E#(I + 4)
        PRINT #1,
        FOR S = 1 TO 60

                IF B#(S, I + 4) < .00001# GOTO 50
                PRINT #1, USING "##    #.##### "; S; B#(S, I);
                PRINT #1, USING "   #.##### "; B#(S, I + 1);
                PRINT #1, USING "   #.##### "; B#(S, I + 2);
                PRINT #1, USING "    #.##### "; B#(S, I + 3);
                PRINT #1, USING "   #.##### "; B#(S, I + 4)
        NEXT S
50      REM space to top of page
        FOR LINECOUNT = 1 TO (66 - S)
```

Table 7.10   Program to Generate Table of Grades of Service Using the Erlang Distribution (continued)

```
            PRINT #1,
            NEXT LINECOUNT
NEXT I
CLOSE #1
END
REM subroutine to compute factorials
100 FOR I = 1 TO 60
            p# = 1
                    FOR j = I TO 1 STEP -1
                    p# = p# * j
                    NEXT j
            FACTORIAL#(I) = p#
            NEXT I
            RETURN
```

to determine equipment or facility size once the concepts involved in the use of the tables are understood. Thus, we will next focus our attention upon the use of traffic tables and their use in the equipment sizing process.

In examining the program listing contained in Table 7.10, note that although the program was written to store output on a file, the program can be easily modified to direct output to a printer. This can be accomplished by removing the statement OPEN "D:ERLANG.DAT" FOR OUTPUT AS #1 and changing all PRINT #1 entries to LPRINT.

Extracts from the execution of the ERLANG.BAS program are listed in Table 7.11. Although the use of the Erlang B formula is normally employed for telephone dimensioning, it can be easily adapted to sizing data communications equipment. As an example of the use of Table 7.11, consider the following situation. Suppose you need to provide customers with a grade of service of 0.1 when the specific traffic intensity is 7.5 erlangs. From Table 7.11, 10 channels or trunks would be required, because the use of the table requires one to interpolate and round to the highest port or channel. Thus, if it was desired to offer a 0.01 grade of service when the traffic intensity was 7 erlangs, you could read down the 7.0 erlang column and determine that between 13 and 14 channels are required. Because you cannot install a fraction of a trunk or channel, 24 channels would be required as we round to the highest channel number.

7.4.4.1   Access Controller Sizing

In applying the Erlang B formula to access controller sizing, an analogy can be made between telephone network trunks and controller ports. Let us assume that a survey of users in a geographic area indicated that during

Table 7.11   Erlang B Distribution Extracts

POISSON   DISTRIBUTION
PROBABILITY ALL PORTS ARE BUSY WHEN CALL ATTEMPTED
WHICH IS THE GRADE OF SERVICE

PORT #   TRAFFIC IN ERLANGS

	5.50	6.00	6.50	7.00	7.50
1	0.99591	0.99752	0.99850	0.99909	0.99945
2	0.97344	0.98265	0.98872	0.99270	0.99530
3	0.91162	0.93803	0.95696	0.97036	0.97974
4	0.79830	0.84880	0.88815	0.91823	0.94085
5	0.64248	0.71494	0.77633	0.82701	0.86794
6	0.47108	0.55432	0.63096	0.69929	0.75856
7	0.31396	0.39369	0.47347	0.55029	0.62184
8	0.19051	0.25602	0.32724	0.40128	0.47536
9	0.10564	0.15276	0.20842	0.27091	0.33803
10	0.05377	0.08392	0.12261	0.16950	0.22359
11	0.02525	0.04262	0.06683	0.09852	0.13776
12	0.01098	0.02009	0.03388	0.05335	0.07924
13	0.00445	0.00882	0.01602	0.02700	0.04266
14	0.00168	0.00362	0.00710	0.01281	0.02156
15	0.00060	0.00140	0.00295	0.00571	0.01026

the BH, normally six personal computer users would be active. This would represent a traffic intensity of 6 erlangs. Suppose we wish to size the access controller to ensure that at most only 1 out of every 100 calls to the device encounters a busy signal. Then our desired grade of service becomes 0.01. From Table 7.11, the 6 erlang column indicates that to obtain a 0.01136 grade of service would require 12 channels or ports, and a 0.00522 grade of service would result if the device had 13 channels. Based upon the preceding data, the access controller would be configured for 13 channels, as illustrated in Figure 7.5.

Based upon a BH traffic intensity of 6 erlangs, 13 dial-in lines, modems, and access controller ports would be required to provide a 0.01 grade of service.

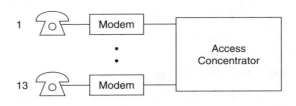

Figure 7.5   Access Controller Sizing

From a practical consideration, the Erlang B formula assumption that lost calls are cleared and traffic not carried vanishes can be interpreted as traffic overflowing one dial-in port is switched to the next port on the telephone company rotary as each dial-in port becomes busy. Thus, traffic overflowing dial-in port n is offered to port n + 1 and the traffic lost by the *n*th dial-in port, e_n, is the total traffic offered to the entire group of dial-in ports multiplied by the probability that all dial-in ports are busy. Thus:

$$e_n = E \times P(T_n, E)$$

where E is the traffic intensity in erlangs and n the number of ports or channels. For the first dial-in port, when n is 1, the proportion of traffic blocked becomes:

$$e_1 = \frac{E}{1 + E}$$

For the second dial-in port, the proportion of traffic lost by that port becomes:

$$e_2 = \frac{E^2/2!}{1 + (E^1/1!) + (E^2/2!)}$$

In general, the proportion of traffic lost by the *n*th port can be expressed as:

$$e_n = \frac{E^n/n!}{1 + (E^1/1!) + \ldots + (E^n/n!)}$$

From the preceding, note that we can analyze the traffic lost by each port on the access controller in the same manner as our previous discussion concerning the computation of lost traffic. In fact, the formulas contained in Table 7.11 for traffic lost and traffic carried by rotary position apply to each rotary position on the access controller. To verify this, let us reduce the complexity of calculations by analyzing the data traffic carried by a group of four dial-in ports connected to a four-channel access controller when a traffic intensity of 3 erlangs is offered to the group.

For the first dial-in port, the proportion of lost traffic becomes:

$$P_1 = \frac{3}{1 + 3} = 0.75$$

The proportion of lost traffic on the first port multiplied by the offered traffic provides the actual amount of lost traffic on Port 1. Thus:

$$e_1 = P_1 \times E = \frac{E}{1+E} \times E = \frac{E^2}{1+E} = 2.25 \text{ erlangs}$$

The total traffic carried on the first access controller port is the difference between the total traffic offered to that port and the traffic that overflows or is lost to the first port. Thus, the total traffic carried by Port 1 is:

$$3 - 2.25 = 0.75 \text{ erlangs}$$

Because we consider the rotary as a device that will pass traffic lost from Port 1 to the remaining ports, we can compute the traffic lost by the second port in a similar manner. Substituting in the formula to determine the proportion of traffic lost, we obtain for the second port:

$$P_2 = \frac{E^2/2!}{1 + \left(E^1/1!\right) + \left(E^2/2!\right)} = 0.5294$$

The amount of traffic lost by the second port, e_2, becomes:

$$e_2 - P_2 \times E = 0.5294 \times 3 = 1.588 \text{ erlangs}$$

The traffic carried by the second port is the difference between the traffic lost by the first port and the traffic lost by the second port, thus:

$$e_1 - e_2 = 2.25 - 1.588 = 0662 \text{ erlangs}$$

A summary of individual port traffic statistics is presented in Table 7.12 for the four-port access controller based upon a traffic intensity of 3 erlangs offered to the device. Note that the computation results contained in Table 7.12 are within a small fraction of the results computed by the traffic analyzer program execution displayed in Table 7.5. The differences between the two can be attributed to the accuracy of a hand calculator versus the use of double precision in the program developed by Gil Held. From Table 7.12, the traffic carried by all four ports totaled 2.3817 erlangs. Because 3 erlangs were offered to the access controller ports, 0.6183 erlangs were lost. The proportion of traffic lost to the group of four ports is e_4/E or 0.6183/3, which is 0.2061. If you examine the ERLANG.DAT

Table 7.12   Individual Port Traffic Statistics

Port	Proportion of Lost Traffic	Amount of Lost Traffic	Traffic Carried
1	0.7500	2.2500	0.7500
2	0.5294	1.5880	0.6620
3	0.3462	1.0380	0.5500
4	0.2061	0.6813	0.4197

file, at the column for a traffic intensity of 3 erlangs and a row of four channels, you will note a similar 0.2061 grade of service. These calculations become extremely important from a financial standpoint if a table lookup results in a device dimensioning that requires an access controller expansion nest to be obtained to service one or only a few ports. Under such circumstances, you may wish to analyze a few of the individual high-order ports to see what the effect of the omission of one or more of those ports will have upon the system.

If data tables are available, the previous individual calculations are greatly simplified. From such tables the grade of service for Channel 1 through Channel 4 with a traffic intensity of 3 erlangs is the proportion of traffic lost to each port. Thus, if tables are available, you only have to multiply the grade of service by the traffic intensity to determine the traffic lost to each port.

7.5   THE POISSON FORMULA

The number of arrivals per unit time at a service location can vary randomly according to one or many probability distributions. The Poisson distribution is a discrete probability distribution because it relates to the number of arrivals per unit time. The general model or formula for this probability distribution is given by the following equation:

$$P(r) = \frac{e^{-n}(r)^r}{r!}$$

where:

r   = number of arrivals
p(r) = probability of arrivals
n   = mean of arrival rate
e   = base of natural logarithms (2.71828)
r!   = r factorial = r x (r − 1) × (r − 2)...3 × 2 × 1

The Poisson distribution corresponds to the assumption of random arrivals, because each arrival is assumed to be independent of other arrivals and also independent of the state of the system. One interesting characteristic of the Poisson distribution is that its mean is equal to its variance. This means that by specifying the mean of the distribution, the entire distribution is specified.

7.5.1 Access Controller Sizing

As an example of the application of the Poisson distribution, let us consider an access controller where user calls arrive at a rate of two per unit period of time. From the Poisson formula, we obtain:

$$P(r) = \frac{2.71828 - 2 \times 2r}{r!}$$

Substituting the values 0, 1, 2, …, 9 for r, we obtain the probability of arrivals listed in Table 7.13, rounding to four decimal places. The probability of arrivals in excess of nine per unit period of time can be computed but is a very small value and was thus eliminated from consideration.

Table 7.13 Poisson Distribution Arrival (Rate of 2 per Unit Time)

Number of Arrivals per Period	Probability
0	0.1358
1	0.2707
2	0.2707
3	0.1805
4	0.0902
5	0.0361
6	0.0120
7	0.0034
8	0.0009
9	0.0002

The probability of the arrival rate being less than or equal to some specific number, n, is the sum of the probabilities of the arrival rate being 0, 1, 2, …, n. This can be expressed mathematically as follows:

$$P(r \le n) = P(R = 0) + P(r = 1) + P(r = 2) + \ldots + P(r = n)$$

This can be expressed in sigma notation (the mathematical shorthand for expressing sums of numbers) as:

$$P(r \le n) = \sum_{r=0}^{n} \frac{e^{-n} \times n^r}{r!}$$

To determine the probability of four or fewer arrivals per unit period of time we obtain:

$$P(r \le n) = \sum_{r=0}^{4} \left[\left(e^{-2} 2^r \right) / r! \right]$$

$$P(r \le 4) = P(r = 0) + P(r = 1) + P(r = 2) + P(r = 3) + P(r = 4)$$

$$= 0.1358 + 0.2707 + 0.2707 + 0.1804 + 0.0902$$

$$= 0.9478$$

From the preceding, almost 95 percent of the time four or fewer calls will arrive at the access controller at the same time, given an arrival rate or traffic intensity of 2. The probability that a number of calls in excess of four arrives during the period is equal to 1 minus the probability of four or fewer calls arriving, which is the grade of service. Thus, the grade of service when a traffic intensity of 2 erlangs is offered to four ports is:

$$P(r > 4) = 1 - P(r < 4) = 1 - 0.9478 = 0.0522$$

If four calls arrive and are being processed, any additional calls are lost and cannot be handled by the access controller. The probability of this occurring is 0.522 for a four-channel access controller, given a traffic intensity of 2 erlangs. In general, when E erlangs of traffic are offered to a service area containing n channels, the probability that the service area will fail to handle the traffic is given by the equation:

$$P(r \ge n) = \sum_{r=n+1}^{\infty} \frac{e^{-e} \times E^r}{r!}$$

Table 7.14 `POISSON.BAS` Program Listing

```
'$DYNAMIC
OPEN "D:POISSON.DAT" FOR OUTPUT AS #1
DIM FACTORIAL#(60)
DIM E#(80), B#(80, 80)
REM E is the offered load in Erlangs
REM S is the number of ports, dial in lines or trunks
REM B(i,j) contains grade of service for port or channel i with
traffic j
GOSUB 100   'compute factorial 1 TO 60
C = 0
FOR I = 5 TO 400 STEP 5
        C = C + 1
        E# = I / 10
        E#(C) = E#
                K = 0
                FOR S = 0 TO 59    'vary port number
                        K = K + 1
                        SX# = S
                        D# = 0
                        FOR X = 0 TO S
                        D1X# = X
                        D# = D# + (E# ^ D1X#) / (FACTORIAL#(X) *
2.71828 ^ E#)
                        NEXT X
                B#(K, C) = 1# - D#
                B#(X, C) = ABS(B#(K, C))
                NEXT S
  NEXT I
  FOR I = 1 TO 80 STEP 5   ' print 16 pages 5 entries per page
        PRINT #1, "                          POISSON  DISTRIBUTION"
        PRINT #1, "         PROBABILITY ALL PORTS ARE BUSY WHEN CALL
ATTEMPTED"
        PRINT #1, "                    WHICH IS THE GRADE OF SERVICE"
        PRINT #1,
        PRINT #1, "PORT #               TRAFFIC IN ERLANGS"
        PRINT #1,
        PRINT #1, USING "     ##.##          ##.##"; E#(I); E#(I + 1);
        PRINT #1, USING "     ##.##"; E#(I + 2);
        PRINT #1, USING "     ##.##          ##.##"; E#(I + 3); E#(I + 4)
        PRINT #1,
        FOR S = 1 TO 60

                IF B#(S, I + 4) < .00001# GOTO 50
                PRINT #1, USING "##  #.##### "; S; B#(S, I);
                PRINT #1, USING "   #.#####"; B#(S, I + 1);
                PRINT #1, USING "   #.#####"; B#(S, I + 2);
                PRINT #1, USING "   #.#####"; B#(S, I + 3);
                PRINT #1, USING "   #.#####"; B#(S, I + 4)
        NEXT S
50      REM space to top of page
        FOR LINECOUNT = 1 TO (66 - S)
        PRINT #1,
        NEXT LINECOUNT
NEXT I
CLOSE #1
END
REM subroutine to compute factorials
100 FOR I = 1 TO 60
        p# = 1
```

Table 7.14 POISSON.BAS Program Listing (continued)

```
            FOR j = I TO 1 STEP -1
            p# = p# * j
            NEXT j
    FACTORIAL#(I) = p#
    NEXT I
    FACTORIAL#(0) = 1#
    RETURN
```

Although commonly known as the Poisson traffic formula, the preceding equation is also known as the Molina equation after the American who first applied it to traffic theory. Because the number of channels or ports is always finite, it is often easier to compute the probability that the number of channels cannot support the traffic intensity in terms of their support. This is because the probability of support plus the probability of not supporting a given traffic intensity must equal unity. Thus, we can rewrite the Molina or Poisson traffic formula as:

$$P(r > n) = 1 - \sum_{r=0}^{n} \frac{e^{-E} \times E^{r}}{r!}$$

To facilitate equipment and facility sizing using the Poisson traffic formula, another program was written in QuickBasic. Table 7.14 contains the program listing of the file labeled POISSON.BAS. This program generates a data file labeled POISSON.DAT, which contains a table of grades of service for traffic intensities ranging from 0.5 to 80 erlangs when lost calls are assumed to be held and thus follow the Poisson distribution.

To execute the POISSON.BAT program using QuickBasic, you must execute that compiler using its /AH option. That option allows dynamic arrays to exceed 64 K.

If you require the generation of a smaller set of tables or wish to alter the increments of traffic in erlangs, both modifications can be easily accomplished. Altering the number of tables requires changing the size of the E# and B# arrays, the statements FOR 1 = 5 TO 800 STEP 5, E# = I/10, and the statement I = 1 TO 160 STEP 5. The FOR I = 5 TO 800 STEP 5 statement controls the number of different traffic loads that will be computed. Because E# is set to I divided by 10, the program listing computes traffic from 0.5 to 80 erlangs in increments of 0.5 erlangs. Thus, you can change the FOR statement or the statement E# = I/10 or both to alter the number of traffic loads and their values. Finally, you must alter the statement I = 1 TO 160 STEP 5 if you change the previously referenced FOR statement. This is because the FOR

I = 1 TO STEP 160 STEP 5 statement controls the printing of output. As included in the POISSON.BAS listing, the FOR I = 1 TO 160 STEP 5 statement results in the printing of 32 pages with 5 entries per page.

Similar to ERLANG.DAT, the file POISSON.DAT is in ASCII format and can be directed to your printer by the command TYPE POISSON.DAT > LPT1: or it can be imported into a word processor that is capable of reading ASCII files. The POISSON.DAT file contains grade of service computations based upon a traffic intensity of 0.5 to 80 erlangs in increments of 0.5 erlangs.

7.5.2 Formula Comparison and Utilization

In order to contrast the difference between Erlang B and Poisson formulas, let us return to the LAN access controller examples previously considered. When 6 erlangs of traffic are offered and it is desired that the grade of service should be 0.01, 13 channels are required when the Erlang B formula is employed. If the Poisson formula is used, an excerpt of one of the tables produced by executing the POISSON.BAS program would appear as indicated in Table 7.15. By using this table, a grade of service of 0.01 for a traffic intensity of 6 erlangs results in a required channel capacity somewhere between 9 and 10. Rounding to the next highest number results in a requirement for 10 channels. Now let us compare

Table 7.15   Poisson Distribution Extracts

POISSON   DISTRIBUTION
PROBABILITY ALL PORTS ARE BUSY WHEN CALL ATTEMPTED
WHICH IS THE GRADE OF SERVICE

PORT #	TRAFFIC IN ERLANGS				
	5.50	6.00	6.50	7.00	7.50
1	0.99591	0.99752	0.99850	0.99909	0.99945
2	0.97344	0.98265	0.98872	0.99270	0.99530
3	0.91162	0.93803	0.95696	0.97036	0.97974
4	0.79830	0.84880	0.88815	0.91823	0.94085
5	0.64248	0.71494	0.77633	0.82701	0.86794
6	0.47108	0.55432	0.63096	0.69929	0.75856
7	0.31396	0.39369	0.47347	0.55029	0.62184
8	0.19051	0.25602	0.32724	0.40128	0.47536
9	0.10564	0.15276	0.20842	0.27091	0.33803
10	0.05377	0.08392	0.12261	0.16950	0.22359
11	0.02525	0.04262	0.06683	0.09852	0.13776
12	0.01098	0.02009	0.03388	0.05335	0.07924
13	0.00445	0.00882	0.01602	0.02700	0.04266
14	0.00168	0.00362	0.00710	0.01281	0.02156
15	0.00060	0.00140	0.00295	0.00571	0.01026

what happens at a higher traffic intensity. For a traffic intensity of 10 erlangs and the same 0.01 grade of service, you will note that 19 channels will be required when the Erlang B formula is used. If the Poisson formula is used you will note that a 0.01 grade of service based upon 10 erlangs of traffic requires between 18 and 19 channels. Rounding to the next highest channel results in the Poisson formula providing the same value as provided through the use of the Erlang B formula.

In general, the Poisson formula produces a more conservative sizing at lower traffic intensities than the Erlang B formula. At higher traffic intensities, the results are reversed. The selection of the appropriate formula depends upon how one visualizes the calling pattern of users of the communications network.

7.5.3 Economic Constraints

In the previous dimensioning exercises, the number of ports or channels selected was based upon a defined level of grade of service. Although we want to size equipment to have a high efficiency and keep network users happy, we must also consider the economics of dimensioning. One method that can be used for economic analysis is the assignment of a dollar value to each erlang-hour of traffic.

For a company such as an ISP, the assignment of a dollar value to each erlang-hour of traffic may be a simple matter. Here, the average revenue per one-hour session could be computed and used as the dollar value assigned to each erlang-hour of traffic. For other organizations, the average hourly usage of employees waiting service could be employed.

As an example of the economics involved in sizing, let us assume lost calls are held, resulting in traffic following a Poisson distribution, and that 7.5 erlangs of traffic can be expected during the BH. From the extract of the execution of the Poisson distribution program, if we desire, initially, to offer a 0.02 grade of service, we will require between 14 and 15 channels. Rounding to the highest number, 15 channels would be selected to provide the desired 0.02 grade of service, which is equivalent to one call in 50 obtaining a busy signal.

LAN access controllers normally consist of a base unit that contains a number of channels or ports and an expansion chassis into which dual-port adapter cards are normally inserted to expand the capacity of the controller. Many times, you may desire to compare the potential revenue loss in comparison to expanding the access controller beyond a certain capacity. As an example of this, consider the data in Table 7.15, which indicates that when the traffic intensity is 5.5 erlangs, a 12-channel access controller would provide an equivalent grade of service. This means that during the BH, 2 erlangs of traffic would be lost and the network designer could then compare the cost of three additional ports on the access controller and additional modems and dial-in lines — if access to the access controller is over the switched network — to the loss of revenue by not being able to service the BH traffic.

7.6 APPLYING THE EQUIPMENT SIZING PROCESS

Many methods are available for end users to obtain data traffic statistics required for sizing communications equipment. Two of the most commonly used methods are based upon user surveys and computer accounting information.

End-user surveys normally require each user to estimate the number of originated calls to a network access point for average and peak traffic situations, as well as the call duration in minutes or fractions of an hour on a daily basis. By accumulating the traffic data for a group of users in a particular geographic area, you then can obtain the traffic that the access controller will be required to support.

Suppose a new application is under consideration at a geographic area currently not served by a firm's data communications network. For this application, ten PCs with the anticipated data traffic denoted in Table 7.16 are to be installed at five small offices in the greater metropolitan area of a city. If each PC user will dial a centrally located LAN access controller, how many dial-in lines, auto-answer modems, and access controller ports are required to provide users with a 98 percent probability of accessing the network upon dialing the LAN access controller? What would happen if a 90 percent probability of access were acceptable?

Table 7.16   Terminal Traffic Survey

Terminal	Calls Originated per Day		Call Duration (minutes)	
	Average	Peak	Average	Peak
A	3	6	15	30
B	2	3	30	60
C	5	5	10	15
D	2	3	15	15
E	2	4	15	30
F	2	4	15	30
G	3	3	15	35
H	4	6	30	30
I	2	3	20	25
J	2	2	15	60

For the 10 PCs listed in Table 7.16, the average daily and peak daily traffic are easily computed. These figures can be obtained through multiplying the number of calls originated each day by the call duration and summing the values for the appropriate average and peak periods. Doing so, you obtain 480 minutes of average daily traffic and 1200 minutes of peak traffic. Dividing those numbers by 60 results in 8 erlangs of average daily traffic and 20 erlangs of peak daily traffic.

Prior to sizing, some additional knowledge and assumptions concerning PC traffic will be necessary. First, from the data contained in most survey forms, information containing BH traffic is nonexistent, although such information is critical for equipment sizing. Although survey forms can be tailored to obtain the number of calls and call duration by specific time intervals, for most users the completion of such precise estimates is a guess at best.

BH traffic can normally be estimated accurately from historical or computer billing and accounting type data or from the use of a network management system that logs usage data. Suppose that the use of one of those sources shows a BH traffic equal to twice the average daily traffic based upon an eight-hour normal operational shift. Then the traffic would be $(8/8) \times 2$ or 2 erlangs, and the BH peak traffic would be $(20/8) \times 2$ or 5 erlangs.

The next process in the sizing procedure is to determine the appropriate sizing formula to apply to the problem. If we assume that users encountering a busy signal will tend to redial the telephone numbers associated with the access controller, the Poisson formula will be applicable. From Table 7.17, the 2 erlang traffic column shows a 0.01656 probability (1.65 percent) of all channels busy for a device containing six channels, 0.05265 for five channels, and 0.14288 for four channels. Thus, to obtain a 98 percent probability of access based upon the daily average traffic would require six channels, and a 90 percent probability of access would require five channels.

If we want to size the equipment based upon the daily peak traffic load, how would sizing differ? We now would use a 5 erlang traffic column contained in the sizing tables. From the table, 11 channels would provide a 0.01369 probability (1.37 percent) of encountering a busy signal and 10 channels would provide a 0.03182 probability. To obtain a 98 percent probability of access statistically would require 11 channels. Because there are only 10 terminals, logic would override statistics and 10 channels or 1 channel per PC would suffice. It should be noted that the statistical approach is based upon a level of traffic that can be generated from an infinite number of computers. Thus, you must also use logic and recognize the limits of the statistical approach when sizing equipment. Because a 0.06809 probability of encountering a busy signal is associated with 9 channels and a 0.13337 probability with 8 channels, 9 channels would be required to obtain a 90 percent probability of access.

Table 7.17   Poisson Distribution Program Extract

```
                    POISSON    DISTRIBUTION
          PROBABILITY ALL PORTS ARE BUSY WHEN CALL ATTEMPTED
                 WHICH IS THE GRADE OF SERVICE

   PORT #                    TRAFFIC IN ERLANGS

          0.50        1.00        1.50        2.00        2.50

      1   0.39347     0.63212     0.77687     0.86466     0.91791
      2   0.09020     0.26424     0.44217     0.59399     0.71270
      3   0.01439     0.08030     0.19115     0.32332     0.45619
      4   0.00175     0.01899     0.06564     0.14288     0.24242
      5   0.00017     0.00366     0.01857     0.05265     0.10882
      6   0.00001     0.00059     0.00446     0.01656     0.04202
      7   0.00000     0.00008     0.00093     0.00453     0.01419
      8   0.00000     0.00001     0.00017     0.00110     0.00425
      9   0.00000     0.00000     0.00003     0.00024     0.00114
     10   0.00000     0.00000     0.00000     0.00005     0.00028
     11   0.00000     0.00000     0.00000     0.00001     0.00006
     12   0.00000     0.00000     0.00000     0.00000     0.00001
```

In Table 7.18, the sizing required for average and peak daily traffic is listed for both 90 percent and 98 percent probability of obtaining access. Note that the difference between supporting the average and peak traffic loads is four channels for both the 90 percent and 98 percent probability of access scenarios, even though peak traffic is two-and-a-half times average traffic.

The last process in the sizing procedure is to determine the number of channels and associated equipment to install. Whether to support the average or peak load will depend upon the critical nature of the application, funds availability, how often peak daily traffic can be expected, and perhaps organizational politics. If peak traffic occurs once per month, we could normally size equipment for the average daily traffic expected. If peak traffic was expected to occur twice each day, we would normally size equipment based upon peak traffic. Traffic between these extremes may require that the final step in the sizing procedure be one of human judgment, incorporating knowledge of economics, and the application into the decision process.

Table 7.18   Channel Requirements Summary

Probability of Access (%)	Daily Average	Traffic Peak
90	5	9
98	6	10

II

LOCAL AREA NETWORK DESIGN TECHNIQUES

This section consists of seven chapters focusing specifically on LAN design techniques using current trends. As noted in Section 1, there are a number of areas where WAN and LAN design techniques overlap and such areas appear with a modicum of reiteration. Also, to provide the network designer with the tools necessary to perform an applicable selection and evaluation of LAN technologies, as well as for the sake of continuity and completeness, there is a fair amount of discussion devoted to conceptual matters.

This section commences with Chapter 8, which focuses on LAN devices. In this chapter, we look at hubs, bridges, switches, routers, brouters, and more. This is followed in succeeding chapters by an overview of topologies and a tutorial on the Ethernet family (the sole survivor of the Ethernet wars), followed by Ethernet performance considerations. In the final chapters, we will consider such technologies as wireless LANs, as well as state-of-the-art issues in the design of internetworks. Some consider the matter of LAN Administration as a design issue, but noting that this requires an up and running LAN to preexist, we view this as rather a modification tool and accordingly omit its discussion.

Finally, there is also a chapter on issues at the network, transport, and application layers and their effect on LAN design, including the possibility of sending delay-sensitive information such as multimedia over an Ethernet network.

8

LOCAL AREA
NETWORK DEVICES

A LAN is a set of locally interconnected devices, connected via the same or different types of media. Because it is important to understand the role of different LAN devices prior to discussing LAN design, this chapter provides an overview of the functionality and capability of different LAN devices. In doing so, we will note that in addition to the usual suspects of PCs, bridges, and routers, we can also consider a mainframe as a LAN station.

8.1 STATIONS AND SEGMENT

Prior to discussing the functionality and capability of LAN devices, we need to cover basic LAN architecture defined by the terms station and segment.

A station represents an active or passive participant on a LAN. As such, a station can represent a PC, mainframe, plotter, or printer. In comparison, a LAN segment is defined to be a set of directly connected stations (via media) with an overall extent of no more than the size of a building or a small group of buildings. This limitation of extent for LANs is in turn a limitation of the technology used in LANs.

Within the limited scope of such a network, within a LAN segment, there are a number of different configurations possible. These are usually referred to as "topologies," which will be discussed in Chapter 9.

In this chapter, we only study the devices that may exist within LAN segments, as well as the devices that may be used to interconnect LAN segments. Note that within LAN segments, one may encounter computers (mainframes, PCs), hubs, and repeaters. On the other hand, bridges, routers, and brouters may be used to interconnect LAN segments. Other devices, such as LAN switches, can be used to interconnect LAN stations, printers, routers, and other network-connected devices. Switches can operate at Layer 2 or higher layers in the OSI Reference Model. Another device worth mentioning is the "gateway," which many persons rightfully confuse with the router as the term gateway is used by most operating systems to reference a router. In this book, we will use the term gateway to reference a device that operates at all seven layers of the OSI stack.

8.2   REPEATERS

As their name suggests, these devices repeat each incoming digital bit, by first actively cleaning up the signal and then regenerating them as they pass on from the repeater. Because a repeater duplicates frames on a bit-by-bit basis, you need to be careful when interconnecting two segments via a repeater. This is because traffic on each segment will then also transport traffic on the other segment. Thus, if Segment A has a 30 percent utilization level and Segment B has a 50 percent utilization level prior to interconnection via a repeater, after the two segments are interconnected, their utilization rates will increase to at least 80 percent. We say *at least* because in an Ethernet environment the probability of collisions increases as the level of utilization increases, resulting in the need for retransmissions, which further increase the level of utilization.

8.3   HUBS

A conventional Ethernet hub represents a repeater. A frame arrives on one port and is broadcast out on all other ports. When a frame is received on one incoming port, it is forwarded out onto all the other ports of the hub, and then subsequently captured by the intended recipient.

As all traffic flows through a shared backplane in the hub, each hub port competes with the others for a slice of the bandwidth. Hence, if the overall bandwidth available in the hub is T bps and N devices are attached to the hub, each device at most obtains a bandwidth of (T/N) bps.

Hubs can be stackable. In other words, two hubs can be interconnected to increase the number of ports available, by cascading ports as illustrated in Figure 8.1. Outgoing information is sent out by the cascading ports to another hub, rather than to an attached device. However, cascading ports tend to be proprietary and nonstandardized, so there is an interoperability problem when one thinks of cascading or stacking hubs from different vendors.

Hubs can be devised at one higher level of sophistication. In this situation, one has intelligent or enterprise hubs or concentrators, wherein

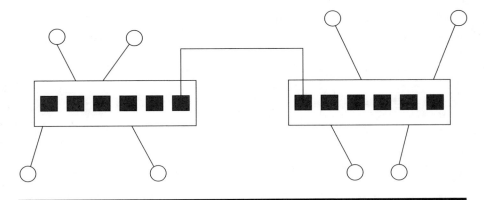

Figure 8.1 The Interconnection of Hubs Forms a Stacked Hub

a given backbone is able to support a number of different modules. For example, the backbone being 100 Base-T, a number of slots are readily available in the hub, wherein cards can be plugged in for 10 Base-T operations or in earlier days possibly Token Ring and Fiber Distributed Data Interface (FDDI). This type of hub has often been called the enterprise hub as well as "network in a box" and "backbone in a box."

However, the basic problem of subdivision of the bandwidth still remains, as *N* devices must share the total available bandwidth.

8.4 BRIDGES

Bridges are devices that operate at the Data Link Layer (DLL). A word of clarification is in order at this point. When a frame arrives at one of the ports of a bridge, the bridge will look at the encapsulated control information related to the DLL to determine where to send the frame. Specifically, the bridge examines the destination MAC address in the Layer 2 header and ignores all upper layer protocol controls.

8.4.1 Types of Bridges

There are a number of different bridge types that are available in the market. As bridges are one of the most critical of LAN technology devices, we will study these in some detail. Bridge types may be itemized as follows:

- Transparent bridges
- Source routing (SR) bridges
- Translational bridges
- Source routing transparent (SRT) bridges

- Source route translational (SR/TL) devices
- Multiport bridges

They may also be further classified by method of operation, such as:

- Cut-through
- Store and forward
- Hybrid

as well as addresses recognized per switch port, i.e:

- Port switches
- Segment switches

These concepts are discussed later in this book when we focus detailed attention on LAN switches.

To understand how these types of bridges work in practice, we must return to the concept of topology that was previously discussed.

As noted, the term topology refers in particular to the geometric configuration of attached devices within a LAN segment. A discussion of this topic occurs in Chapter 9, but for now we can note that bridges can be used to either conjoin disparate LAN segments or to break up a large, bulky LAN segment into smaller, more manageable segments, especially in the face of poor performance within the parent segment. When we consider the configuration of a number of different LAN segments connected by two-port bridges, we obtain a different flavor of topology, and will refer to this as bridged topology.

8.4.2 The Learning Bridge

The notion of associating a graph with a set of bridged LAN segments, i.e., with a bridged topology, is central to the notion of a learning bridge.

Learning bridges work as follows. Bridges learn addresses based upon source address, but switch frames based upon destination address:

- The bridge maintains an internal table of `<Port #, MAC addresses, time entry>`.
- When a frame arrives via one of the bridge ports (say p1), from a certain source MAC address (ma1), to a destination address (ta1), the bridge first learns the source addresses in incoming frames, but they forward based on destination addresses. The bridge's learning and frame switching behavior occur as follows:

■ Learning
  - If <p1, sa1> exists in the table, do nothing
  - If no entry in table for sa1, add <p1, sa1>
  - If <p2, sa1> exists, replace it with <p1, sa1>

■ Switching
  - Look at the destination address ta1.
  - No entry in table for ta1 → "flood" the frame out all ports other than the incoming port where the frame was received.
  - Unique entry in table (not the incoming port p1) → "forward" out that port.
  - <p1, ta1> in table then "filter" the frame, in effect discarding it.

Thus, the operation of a bridge is often referred to as Three F's—flooding, filtering, and forwarding.

Eventually, a full table of <Port #, MAC address> is constructed, and the bridge has learned how to directly handle all source-target routing decisions between the connected LAN segments. It is then able to correctly forward Layer 2 information between the LAN segments it connects.

Figure 8.2 illustrates an example of the construction of a bridge port and address table. Initially, the bridge is powered on and its port and address table is empty as shown in the lower left portion of the illustration. Next, it was assumed that a frame whose source address is A and whose destination address is E was received on Port 0. Because the bridge notes Source Address A occurred on Port 0, it can update its port and address table. However, because it does not know where the destination address resides, it floods the frame, sending it out all ports other than the port it arrived on. The update of the port address table is shown in the middle of the lower portion of Figure 8.2. Finally in our example, a frame is transmitted from Station G connected to Port 3 on the bridge to Station A. The bridge consults its port and address table and knows that Address A is associated with Port 0, so it switches the frame to that port. In addition, the bridge determines that Source Address G is associated with Port 3 and updates its port and address table as illustrated in the lower right portion of Figure 8.2.

Realize that there are changes possible in the neighborhood of the bridge, e.g., a device is unhooked from a particular LAN segment and plugged into another LAN segment attached to the bridge. The bridge learns about this as well. Similarly, if a device has just been unplugged from a certain LAN segment (and not reattached elsewhere), the bridge learns this as well by discovering the source address of frames flowing onto the new port to which that device is now connected. Because bridge

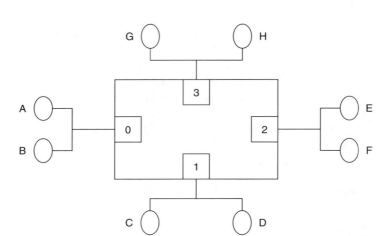

Figure 8.2   Constructing the Port and Address Table

memory is finite and the searching for addresses takes time, a third metric is used in bridge tables. That entry is a timestamp, enabling the bridge to periodically discard old entries.

As noted earlier, bridges operate at Layer 2 (DLL) of the OSI Reference model. In other words, bridges examine only Layer 2 encapsulations within the information frames (which could have encapsulations from each of the other higher layers in the OSI Reference Model) and makes decisions based on the information therein, in particular the source and destination MAC addresses. Transparent bridges make no changes to the content or format of the frames received, nor any further encapsulation. Each frame is repeated, bit by bit, onto the port where the target destination address is assumed to reside. The bridge acquires or learns addresses and performs routing intelligently as previously described. If the actual destination is several bridges away from the source, then that route, via intermediate bridges, needs to be learned as well.

8.4.3 Translational Bridges

Because different types of LANs (various Ethernet family members, 802.3, Token Ring, and ATM) employ different frame formats and methods of control at the DLL, an intelligent bridge must be sensitive enough to be able to distinguish between the different possible controls within incoming frames and handle each situation appropriately. Such a bridge is called a "translational bridge" or "intelligent bridge." On the other hand, a learning bridge connecting two LANs using the same DLL protocol is called a transparent bridge.

8.4.4 Bridged Topologies

We now return again to the notion of bridged topologies, where different LAN segments are connected by multiple bridges. Herein we find one important problem that is associated with the learning approach introduced above for bridges to learn routing information.

This problem occurs when there is a loop in the bridged topology. In this situation, we see that there will be a repeated thrashing within the system, resulting in packets going around the loop indefinitely. Once this was noticed, a solution was quickly found as well. When there are loops in the graph associated with a bridged topology graph, an algorithm popularly known as Kruskal's algorithm was deployed to create a structure without a loop, referred to as an MST. Kruskal's algorithm works as follows:

1. Sort the edges of the bridged topology graph (G) into increasing order by weight or length.
2. Construct a subgraph (S) of G and initially set S to be the empty graph.
3. For each edge (e) in sorted order, if the endpoints of the edges (e) are disconnected in S, add them to S.
4. Continue as long as possible and set S equal to the MST of G.

An example of Kruskal's algorithm is illustrated in Figure 8.3. In this example, we see a network graph evolving into an MST as follows:

1. Sort the edges in increasing order of weight
2. Initial S = φ
3. S = <A, C>
4. S = <A,C> + <B,D>
5. S = <A,C> + <B,D> + <C,B>

This is now the MST shown on the right side of Figure 8.3.

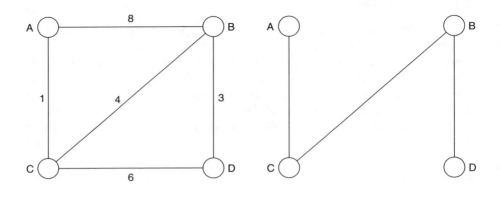

Figure 8.3 An Example of the Use of Kruskal's Algorithm

8.4.5 SR Bridge

Another type of bridge, one that used to be commonly encountered in earlier days, is called the SR bridge. SR bridges connect disparate networks. A source station sends explorer packets into the network. As an explorer packet reaches the target, each ring and bridge en route has placed its stamp on the explorer. When all explorers arrive (at the target), the destination ring sensibly chooses one source route, depending on its requirement. Because Token Ring LANs are rapidly being phased out in favor of Ethernet LANs, these SR bridges are on the way out.

8.4.6 SRT Bridges

A third category of bridge is the SRT bridge. This type of bridge is usually used to interconnect different types of LANs, such as a Token Ring and an Ethernet LAN. On the Token Ring side, they examine incoming frames for the presence of a specific control field called the routing information indicator (RII). If an RII is present, then the frame is source routed. If not, the SRT bridge transparently bridges the frame.

8.4.7 SR/TL Bridges

Another type of bridge is the SR/TL. This type of bridge allows devices in an Ethernet domain to communicate successfully with devices in an SR domain via a Cisco router, which performs the source route translation. In this sense, interconnected extended LANs attached via SR bridges are connected with extended LANs attached via transparent bridges.

8.4.8   Remote Bridges

A remote bridge consists of LAN and WAN ports, in effect providing a mechanism to interconnect two geographically separated LANs via a WAN. Assuming a remote bridge connected to LAN A, the bridge encapsulates all Layer 2 information into a tunnel. The bridge also wraps logical link control (LLC) and MAC information into another protocol for transmission over a WAN using a point-to-point link. On the sending side, users generate complete information packets. The remote bridge simply tunnels the DLL header or trailer into a Point-to-Point Protocol (PPP) that is transmitted across the WAN. On the receiving side, the remote bridge at the opposite end of the WAN connection strips off all tunneling Layer 3 controls, processes the data link information, and passes the information to the destination MAC address transported in the frame previously tunneled via the WAN. For this reason, remote bridges are sometimes called half bridges with two remote bridges and one point-to-point WAN link forming a unit.

8.5   LAN SWITCHES

LAN switches were invented primarily to address the problem of bandwidth sharing between device connections attached to a traditional hub.

8.5.1   The Basic Premise

To redress the problem of bandwidth sharing on conventional hubs, developers introduced multiport bridges that are more popularly referred to as LAN switches. The basic premise behind LAN switches is the notion of dedicated backplane bandwidth for individual connections. As previously noted, in a hub, if the backplane bandwidth is T and N connections are in place, the bandwidth per connection is on the average T/N. By using a switching facility, each connection is able to obtain on a frame-by-frame basis the full T bandwidth via intelligent switching.

The LAN switch evolved from the basic design of a PBX. Through the switching of input to output, one is able to provide each of multiple users with a bandwidth equal to that of the backbone itself. This is sometimes called parallel networking. The switch is able to create connections between two attached Ethernet devices with a latency of less than 40 ms, overcoming the one-at-a-time broadcast limitation of conventional hubs. As you will note, a fabric is created that has N × N cross-points (where the switch has N ports). In this manner, N/2 cross-connections are simultaneously possible, so if the overall available bandwidth is T bps, the utilized throughput is actually (N/2 × T) bps at any point in time.

8.5.2 Delay Timing

There are two associated delays you need to be concerned about when considering LAN switches:

1. Latency — this is the time required to convert an incoming packet and send it out the switch.
2. Delay Time — when many connections share a common target destination, what is the switch's blocking time? (Note: Latency is typically ~ 40 ms per packet.)

Delay time for a 1526-byte maximum length Ethernet frame is typically ~ 1.22 ms. This is arrived at as follows. Consider a maximum length Ethernet frame of 1526 bytes consisting of 1500 bytes of data and 26 bytes of control overhead. At a 10 Mbps operating rate, each bit time is 1/10 exp(7) or 100 nanoseconds (ns). For a 1526-byte frame, the minimum delay if one frame preceded it in attempting to be routed to a common destination is:

$$1526 \text{ bytes} \times 8 \text{ (bits/byte)} \times 100 \text{ ns/bit} = 1.22 \text{ ms}$$

8.5.3 Types of LAN Switches

There are four types of LAN switches:

1. Cut-through
2. Store and forward
3. Hybrid
4. Fragment-free

With cut-through switches, only the destination MAC address in incoming frames is examined, and based on that address only a forwarding decision is made. No other checks occur. Thus, a cut-through switch has the lowest delay and should be considered for supporting real-time applications, such as VoIP and streaming media.

With a store and forward switch, the entire frame is copied into the switch's internal memory, examined for occurrence of any errors, then sent out the right port. Because the entire frame is stored and the frame is variable in length, the delay is also variable. As errored frames are removed by the destination device on a LAN, the necessity of such error checking by a LAN switch has been questioned. However, the filtering capacity should be more useful to route protocols carried in frames to destination ports more easily than by frame destination address. Especially if one has hundreds and thousands of devices attached to a large switch.

Hybrid switches represent a combination of cut-through and store and forward switches. They work as cut-through switches until a certain level or threshold of errors is reached, at which point they revert to performing as store and forward switches. This means that the efficiency of these types of switches is also variable.

The major advantage of a hybrid switch is its minimal latency when error rates are low and it becomes a store and forward when error rates rise, allowing it to discard frames when error rates get high.

A fourth type of switch, fragment-free, examines the first 64 bytes of incoming frames for any errors. If none are found, they push out the entire frame (without storing it), using the belief that most errors are likely to occur there in the first 64 bytes. Because fragment-free switches have a slightly longer delay than cut-through switches, but the delay is uniform, they can usually be used for VoIP and streaming media applications.

8.5.4   Switch Design Methods

LAN switches can be designed in three possible ways:

1. Shared memory
2. Matrix
3. Bus architecture

Shared memory switches store all incoming packets into a shared memory buffer. This buffer is common to all switch ports (I/O). The information is then sent out the correct port. This introduces a maximum delay in processing and can be used for connections not so delay sensitive.

Matrix switches have an internal grid wherein input ports and output ports cross each other. When a packet comes in on an input port and is detected there, the MAC address is compared to the lookup table to find the appropriate output port. The switch then makes the appropriate connection on the grid where these two ports intersect. This is the fastest of the three designs and is useful for such delay-sensitive applications such a streaming multimedia and voice connections.

Continuing our brief discussion of architecture, the bus architecture uses a common internal bus transmission path that is shared by all ports using time division multiple access (TDMA). In this situation, the switch has a dedicated memory buffer for each port, as well as an application-specific integrated circuit (ASIC) for internal bus access. This is a hybrid between shared memory and matrix architecture and provides a performance speed between the two.

8.5.5 Network Utilization

Finally, before we leave the discussion on LAN switches, let us understand the ways in which LAN switches can be used:

- Network redistribution
- Server segmentation
- Backbone

Network redistribution occurs when a LAN switch is used to directly connect via "fat pipes" to workstations that have significantly high levels of traffic. Fat pipes refers to a group of switch ports that function as a single entity, providing a higher bandwidth than obtainable by a single connection between a switch and a networking device. When busy workstations or servers are connected to a switch, the connection usually occurs on a fat pipe. By comparison, when connecting less busy devices to a switch, they are normally connected to conventional switch ports.

Server segmentation occurs by adjoining busy servers directly to switch ports via fat pipes. In this sense, network and server segmentation are similar.

Finally, LAN switches can be used via a backbone at higher speeds servicing lower speed conventional hubs and servers as well. Note that in this regard it is also possible to replace the switch by a conventional high-speed hub. Whereas that decision cuts down the available bandwidth, it represents a less expensive option.

Switches can be further subdivided as follows:

- Port switches
- Segment switches

The former associates a unique address with any given port, whereas in the latter, one can attach a full traditional LAN segment to a switch port, which gives a greater level of access for the stations on the segment to servers attached directly to the LAN switch.

8.6 ROUTERS

In this section, we will turn our attention to the router. As previously mentioned in this chapter, when we discuss the router we are discussing a device that operates at Layer 3 of the OSI Reference Model and which is referred to as a gateway in most operating systems.

8.6.1 Routers in Relation to LANs

There are many potential problems associated with learning bridges and their associated routing behaviors. As the number of interconnected LANs rises, with an attendant increase in the associated number of workstations therein, performance degrades considerably. Performance issues are considered in detail in Chapter 11. Routers do not make decisions based on MAC addresses. Instead, they look to controls (control information) within the Layer 3 (network layer in the OSI stack) headers within the information frame. They must be intelligent enough to use the network layer addressing contained in incoming frames. If an incoming frame refers to one of the router's interfaces, the router processes it. If it does not, the router ignores that frame. This is called "nonpromiscuous operation." Based on the information there, including the target address, the router will make an intelligent decision about what the next hop for that packet should be.

Let us take a quick break at this point to discuss routers in relation to our agenda of LANs.

No LAN is an island. There is the need to communicate with LANs across an internet. When network sizes grow large, it becomes difficult to bridge all data across a WAN. In particular, bridging tables grow excessively large and performance degrades considerably. For this reason, routers were conceived to operate at the network layer and implement more sophisticated switching and routing algorithms than Ethernet bridges.

The earliest routers were referred to as gateways, as they provided a gateway for stations to have their transmissions flow off the LAN. Although Microsoft Windows and other operating systems still use the term gateway, they are really referencing the address of a router in their configuration screens. Concerning addressing, routes operate at Layer 3. Because frames flow on a LAN using Layer 2 addressing, you need to configure workstations with their Layer 3 addresses. Routers and workstations use ARP to learn the Layer 2 addresses of other devices on a LAN, because they cannot directly transmit to one another via the use of Layer 3 addresses.

As a result, routers and their behavior become critical to any treatment of the interoperability of LANs and WANs. Accordingly, we introduce router behavior in this chapter as well.

8.6.2 Router Behavior

There are two aspects to router operation we need to be concerned about. Those aspects are:

1. Deploying an appropriate addressing scheme and control information within the network layer encapsulation.
2. Using suitable techniques to route information packets across the internet. In particular, this includes determining the next hop.

Protocols at the level of (1) above are commonly called "route protocols," whereas those at the level of (2) above are called "routing protocols."

Examples of route protocols include IP, Internet packet exchange (IPX), AppleTalk® network system, Digital Equipment Corporation's DEC-Net, and Xerox® network system (XNS). Examples of routing protocols include Routing Information Protocol (RIP), Open Shortest Path First (OSPF), Intermediate System 2 Information System (IS-IS), and Real-Time Management Protocol (RTMP).

Route protocols function as the name and address information on letters to be delivered by the International Postal System. In comparison, routing protocols determine the specific route that an individual letter will traverse to get to the target name and address.

Here is an example:

S Ravi Jagannathan
2/28 Old South Head Road
Rose Bay NSW 2135
AUSTRALIA

The above identifies the (target) address for a delivery and is posted say in Macon, GA (USA). Whereas the destination is fixed, the path taken is varied. The letter could go:

- Macon→Chicago→Seattle→Sydney
- Macon→Houston→Los Angeles→Sydney
- Macon→Houston→San Francisco→Sydney
- Macon→San Francisco→Honolulu→Auckland→Sydney

All of the above routes are valid. The routing protocol used might select one (or two) route(s) as being the best. Also, if the second route was deemed the best, but the Los Angeles post office was not operational at some point in time, there is provision for rerouting through another route to Sydney. In this context, LAN addressing and routing may be seen to be the delivery of mail to the correct pigeonhole within the target organization's mail room.

8.6.3 Types of Routers

Routers can be classified as:

- Uniprotocol
- Multiprotocol (intelligent)
- Protocol independent (PI)
- Nonroutable protocol support

Uniprotocol routers support and understand only one route protocol and one routing protocol. Note again that when we say protocol, we refer to control information encapsulated at the corresponding layer. For routers, this is Layer 3, the network layer, and serves to explain the behavior of both uniprotocol and multiprotocol routers.

Multiprotocol routers are able to intelligently identify which protocols are in fact being referenced and will accordingly adjust their behavior.

PI routers do not have to or does not look at control information from protocols, particularly from protocols that do not support the concept of network addresses. PI routers function as a sophisticated learning bridge and actively learn which devices are on which network by examining source addresses in frames coming from connected networks. Therefore, both routable and nonroutable protocols can be serviced by PI routers.

Again, consider the problem of devices moving around on LAN segments or the case when a LAN segment is dissected into two subsegments. All network addresses of the devices moved will change. Determining the new addresses imposes a significant administrative burden. This problem is alleviated to an extent by the use of a PI router.

Support for Nonroutable Protocols: Although we tend to primarily think of the Internet when we discuss the movement of data from one LAN to another, there are many older networks that have difficulty being connected in a modern communications environment. These networks were developed prior to the need to interconnect separate networks and are nonaddressable. Because they do not have distinct LAN addresses, they are often referred to as nonroutable networks and their protocols are referred to as nonroutable protocols. One common example of a nonaddressable network protocol is IBM's System Network Architecture (SNA).

Assume that a host using SNA protocol needs to communicate across a network with another host sitting on a LAN system. Because SNA is intrinsically nonroutable, it becomes necessary to employ one of three mechanisms to enable communication:

1. Tunnel (encapsulate) the SNA information (and vice versa)
2. Convert the protocols
3. Use PI routers

It is now time to look at some of the outer internals of route protocols as well as at routing protocols.

8.6.4 Route Protocols

In an IP environment, the addressing scheme used by IPv4 uses 32-bit binary numbers for source and target addresses. Therefore, up to

2,294,967,296 devices can theoretically be addressed. Although this may seem to represent a large number, with the Internet PC population explosion, onslaught of mobile phones, and the growing number of users, this is still not quite enough. Also, every interface on a router or gateway needs a distinct IP address. So a new standard was developed referred to as IPng (IP next generation), better known as Ipv6, with 16 byte (128-bit) addresses and is actively being investigated and promoted as we go to press.

Note that IPv4 subdivides the 32-bit addresses into network and host parts. Typically, organizations can purchase a certain number of appropriate network addresses and then allocate the associated host addresses to devices within their infrastructure.

8.6.5 Routing Protocols

In this domain, there are two different types of algorithms:

1. Vector Distance routing algorithms
2. Link State algorithms

8.6.5.1 Vector Distance

A Vector Distance algorithm maintains information about the next best hop from a given network device. At any point in time, a router will have attached to it a list of `<destination addresses, next hop, distance>`. Periodically, routers exchange control information about the cost of reaching the destination addresses from that router. Consider two routers, R1 and R2. Consider a target address, A. R2 maintains information about the cost, C2, of reaching A via R. Suppose R1 advertises a cost, C1, of reaching A from R1. Then, R2 constructs the number Dist(R1, R2) + C1:

> If [C3 = Dist(R1, R2) + C1] < C2, then R2 will replace the entry
> `<A, R, C2>` by `<A, R1, C3>`

It can be shown that this process, over all routers in the network, actually stabilizes in time and yields a suitable method of managing the routing of information in a packet switched network. It is outside the scope of this book to discuss the pros and cons of this method, including its performance. We will merely compare this approach, however, with the other generic approach mentioned above, namely Link State routing.

8.6.5.2 Link State

Link State protocols differ from Vector Distance in many respects. Unlike the situation wherein a table of information is exchanged periodically, Link State protocols will advertise network changes if and only if there is a change in the network scenario. At that time, the only information sent out is whether a particular link is up or down. Each router actually maintains a complete route to the destination node, i.e., it appreciates the entire topology of the network. When a particular link goes down, another route is automatically computed. Finally, unlike Vector Distance, where the metric used for comparing routes is the hop distance, for Link State, the metric can be, for example, link delay, capacity, or reliability. Also, Link State supports load balancing between different routes to the destination.

Link State can be summarized as follows:

■ Each router transmits a greeting to its neighbors.
■ Each router constructs a Link State Packet (LSP), which is a list of neighbors and the cost to them.
■ All routers in the network receive LSPs from all other routers. This gives a global visibility and a map of the network's complete topology.
■ Based on the LSPs, each router calculates the preferred routes to the destination.

8.7 BROUTERS

As their name suggests, these devices are a cross or hybrid between bridges and routers. A brouter first examines incoming information for the presence of recognizable network layer protocols (control information). If one is visible, the brouter behaves like a router and forwards the packet via its route/routing protocols. If no Layer 3 protocols are recognized, the brouter will bridge the frame appropriately.

8.8 GATEWAYS

Gateways are devices that provide protocol translation support at all seven layers of the OSI protocol stack (or at all five layers of the TCP/IP protocol stack). Typically, gateways have been associated with the provision of application-level translation support, for example, e-mail.

An organization may use Lotus CC:Mail within its intranet. This protocol has its own endemic controls at the application layer (as well as below), so bridges and routers will be unable to readily translate them.

Here is where the gateway concept comes into play. It can connect the CC:mail information for the intranet into the Standard Mail Transfer Protocol (SMTP) e-mail protocol used on the Internet and vice versa, thus providing connectivity. Additionally, gateways provide connectivity to a mainframe or via a remote packet switched network. Critically, understand that gateways work at all seven layers of the OSI Reference Model and for LAN-WAN traffic (rather than just LAN-LAN traffic). This is their distinguishing characteristic. Because of this, gateways tend to be slower than routers and more costly to procure. Finally, if a gateway supports translation between two given protocols, it is called a Two-In-One (2 in 1) gateway. Support for N protocols yields an N-In-One (N in 1) gateway.

Gateways also interface mainframes and user workstations on local and remote LANs. This can occur in the following ways:

- Mainframe to Ethernet LAN
- SNA ↔ LAN Traffic via a software development life cycle SDLC-LAN card
- Over a packet WAN using an X.25 gateway
- 3278/9 coaxial connectivity
- 3172 interconnect controllers

8.9   NETWORK INTERFACE CARDS

The network interface card (NIC) can be viewed as the actual interface between workstations, servers, clients, and the shared media of the network. NICs implement the MAC in software. Therefore, they work at the DLL, using MAC addresses to control the flow of information for devices onto the LAN media and vice versa. Thus, it is safe to say that the NIC is the physical interface between devices and the LAN media. The NIC connects the device and the network media, for example, on coaxial cable by way of a transceiver cable (sometimes called AUI — attachment unit interface), which ends on the media side with a T-connector joining the cable to the media via the transceiver (sometimes called MAU — medium attachment unit). NICs can be plug and play. For example, remove a 10BASE-T NIC card from the system expansion slot of a PC and replace it with a 100BASE-T NIC card and you have a Fast Ethernet workstation. The NIC adds the preamble to frames for synchronization, which explains why minimum and maximum LAN frames are eight bytes less inside a workstation than on a LAN.

8.10   FILE SERVERS

When the LAN concept gained popularity, network software vendors introduced the notion of client-server computing, wherein a server provides services used by client nodes. Clients request services and servers oblige with a response. Over time, four different types of servers emerged in the industry:

1. Print servers
2. Application servers
3. Remote Access Servers (RAS)
4. Communications servers

Print servers control shared printers and manage the printing requirements of a number of client workstations. Typically, a print server consists of a device driver that also supports multiple access in some form.

Application servers control a repository of shared applications and provide client hosts with a copy of the applications for localized usage.

The RAS represents a hybrid device consisting of a modem pool and a router. It supports serial communication in terms of analog modems, Integrated Services Digital Networks (ISDNs), and T1 lines. Incoming communications are validated for security purposes prior to getting access to the network serviced by the RAS.

A communications server is similar in nature to a RAS, but is intended solely for outbound traffic. It may not support inbound traffic and functions as a replacement for users with individual modems. Sometimes they are called "modem poolers."

9

LOCAL AREA NETWORK TOPOLOGIES

In this chapter, we turn our attention to the topology of different LAN solutions and the overall structure or architecture of popular LAN solutions. Key topologies examined in this chapter include the loop, bus, tree, star, and mixed (hybrid) topologies. Ring topologies are not treated, as they are on the way out in relation to current networking trends. By topology, we mean the geometry and geography of interconnected LAN stations and segments. In studying these topics, we will also have an opportunity to briefly examine the different media that connect LAN devices. We will also look at different encoding techniques, as well as various MAC protocols.

9.1 INTRODUCTION

Figure 9.1 depicts the appearance of popular LAN topologies, including:

- Loop
- Bus
- Token bus
- Tree
- Star

The loop requires a controller that polls each station on the network. Although this action ensures no two stations can communicate at the same time, alleviating the possibility of a collision, the polling represents additional overhead.

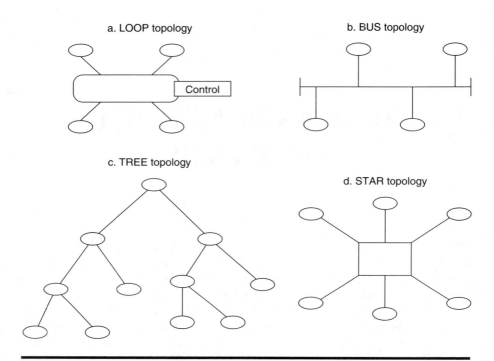

Figure 9.1   LAN Topologies

Bus and tree topologies use a multipoint medium with a shared access protocol. The bus is a special case of the tree with only one trunk and no branches. Transmission on a bus or tree can be sent by any station and is received by all other stations.

The star configuration connects a central hub (or cascaded series of hubs) to a number of stations. As with a bus or tree, a transmission from any station can be received by all others. Details of these topologies are presented in Section 9.2.

Note that LAN devices are interconnected using different types of media, such as copper coaxial, twisted pair, over the air or wireless, and various types of fiber optics. Signals are sent on the media according to various encoding techniques developed to optimize the use of different types of media.

Currently available media are:

- Twisted pair (shielded and unshielded)
- Coaxial cable (Coax)
- Optical fiber
- Wireless

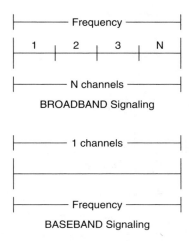

Figure 9.2   Broadband and Baseband Signaling

Each type of media is discussed in more detail later in this chapter and in Chapter 10.

Regarding encoding, a word here will be appropriate and useful from a helicopter view standpoint. There are two basic approaches to encoding signals for transmission:

1. Broadband (analog)
2. Baseband (digital)

Figure 9.2 illustrates an example of broadband and baseband encoding. Broadband represents an analog signaling method that subdivides available bandwidth into a number of separate subchannels, separated by guard bands. Within a subchannel, analog signals carry information.

Under broadband signaling, one of the following modulation techniques is used:

■ Frequency shift keying
■ Amplitude shift keying
■ Phase shift keying

Broadband can use individual subchannels to carry voice, video, and data. This transmission method applies to both bus and tree topologies and can support distances of up to tens of thousands of kilometers. The guard bands are unused blocks of frequency that enable analog signals to slightly drift by frequency without one channel adversely effecting another.

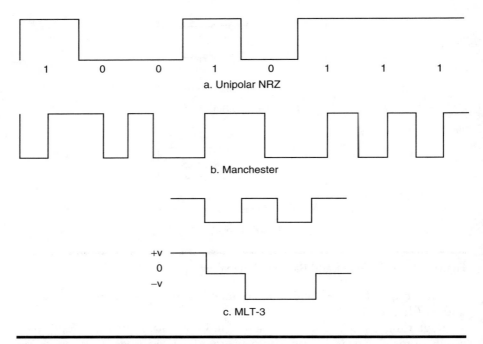

Figure 9.3 Common Digital Encoding Methods

In comparison, baseband, which represents a digital signaling technique that is bidirectional, uses the entire bandwidth of the transmission media as one channel. In this regard, there are a number of methods to actually encode the data. We quickly mention these now, as they are important to LAN capacity, a subsequent consideration in this chapter:

■ Unipolar NRZ (Nonreturn to Zero)
■ Manchester
■ Differential Manchester
■ MLT-3
■ 4B5B
■ 8B6T
■ 8B10B

In Chapter 10, we provide a detailed discussion of encoding techniques.

Figure 9.3 compares three common digital encoding techniques. The first encoding technique, unipolar nonreturn to zero, represents the internal signaling used by computers, terminals, and communications devices. You can only transmit relatively short distances using this signaling method. In addition, clocking is required to distinguish a sequence of set or nonset bits. The other encoding methods make it possible to distinguish one set or nonset bit from a sequence of such bits without clocking, therefore providing a more economical transmission method.

9.2 KEY TOPOLOGIES

In this section, we introduce and discuss the following topologies:

- Loop
- Bus
- Tree
- Star
- Hybrid

9.2.1 The Loop Topology

The loop topology is considered by many to be extinct, but it introduces some important ideas worth noting. A central controller polls the client stations and allocates time to each individual connection. This approach allows for minimum intelligence in the stations (terminals). One problem with this topology is that the controller may not be able to intelligently manage rogue (greedy) connections that dominate the controller. We will return to this matter again later in this chapter when we discuss the Demand Priority Media Access (DPMA) protocol. Two additional problems concern the controller and polling overhead. If the controller crashes, so does the entire loop network. Because the controller must poll each station, this introduces additional overhead. Although developed by IBM during the 1970s, this topology was primarily limited to interconnecting bank terminals and did not achieve acceptance outside the financial community.

9.2.2 The Bus Topology

The bus topology consists of a linear segment of media, to which stations are attached. If instead of stations, other segments are attached, then we obtain a tree topology.

The bus topology is primarily a broadcast mechanism, where outgoing messages from one station are relayed to all stations on the segment in the hope that the desired recipient will recognize and accept it. Stations that are not the nominated recipient simply ignore the frame, with the term frame referencing data padded with control information.

The bus topology introduces two problems:

1. How does the intended recipient correctly recognize that the data is indeed meant for them? On a bus there is no need to pass information, unless the station is a repeater or bridge. A regular station copies each frame on the media, examines its destination address, and either reads it or discards it.

2. If two stations, S_1 and S_2, transmit nearly simultaneously, then there can be a collision on the segment, which basically corrupts the data prior to its reaching the proper destination. How is this problem to be rectified?

The first problem is taken care of by affixing to the frame a control field with the source and destination MAC addresses. Stations attached to the network are able to determine if the destination address matches their own. The second problem is managed using a MAC protocol.

There are six types of MAC protocols. These are included here for the sake of completeness and continuity of the discussion and because they affect network performance and frame rates, as further discussed in Chapter 11. They are also useful when designing, for instance, 100VG-Any-LANs. MAC protocols include:

- Carrier Sense Multiple Access with Collision Detection (CSMA/CD)
- Carrier Sense Multiple Access with Collision Avoidance (CSMA/CA)
- Token passing
- Connection-oriented
- DPMA
- Round robin polling

9.2.3 The Tree Topology

This is an elaborate bus. Instead of individual stations, whole segments are attached to the shared media. Again a broadcast approach is used; any message from a node of the tree traverses up the tree to the root node (head end) and then down to all other nodes to be absorbed by the intended recipient. From a network design viewpoint, this results in two propagation delays when one station transmits to another on the same branch. It is implicitly understood that the tree has no closed loops to prevent endless perambulation.

9.2.4 Star Topology

In this approach, a star is formed with a central controller, with radiating rays going outward reaching to stations. Figure 9.4 illustrates an example of a star topology.

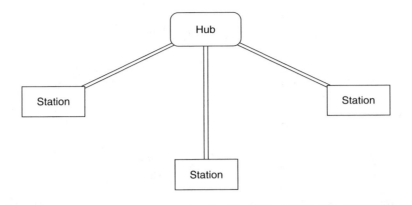

Figure 9.4   Star Topology

A key advantage of star topology deployment is that it allows you to use existing unshielded twisted pair cable, which exists in most buildings as simple telephone wire, running from the telecommunications closets to each office and workstation. Thus, there is usually no cable installation cost. On the other hand, coaxial cable has to be pulled to each office, which can be problematic when existing conduits get crowded. Also it can be expensive to run coax cable to every room in the building. Furthermore, if there is a relocation of equipment to an area where there is no prior coverage, it becomes very costly to enhance the coax coverage.

Figure 9.4 depicts the star topology coverage that is usually the preferred mode of coverage. The central hub is connected to all stations via two links (send and receive). The hub is not intelligent. When a station transmits, the hub repeats the signal on all outgoing links. The star is really a bus logically and the same problems accrue. That is, if two or more stations transmit at the same time, there is a collision. So some "smart" MAC protocol must be used. Also, because of the high data rate and the attendant poor transmission properties of an unshielded twisted pair (UTP), each link is limited to about 100 m.

Figure 9.5 shows a hierarchical configuration of cascading hubs. There is a top-level hub and lower levels of middle-level hubs. Each hub is attached in a tree fashion to either other hubs or to LAN stations. The tree mechanism of managing frames applies. When a signal arrives at the top-level hub, it is repeated onto all outgoing links on that hub, including links to other (middle-level) hubs, as well as stations connected to the top-level hub. At intermediate hubs, signals coming from above are cascaded to all links going down and signals from below are in turn sent on the path to the top-level hub, which then processes them as usual.

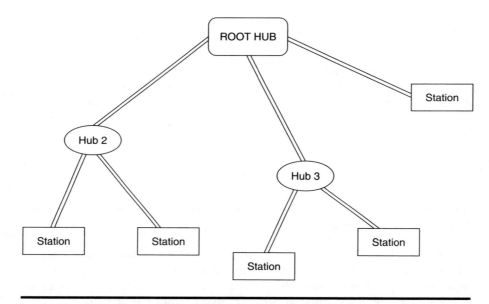

Figure 9.5 Hierarchical Configuration of Cascading Hubs

Initially, hubs were deployed that supported a data rate of 1 Mbps. There quickly followed many products at 10 Mbps that were interoperable with baseband coaxial cable systems, if the transceiver were changed. More recently, there has been considerable interest in hubs operating at 100 Mbps and even 1 and 10 Gbps.

9.2.5 High Data Rates and the Associated Problems

At high data rates, a number of hitherto acceptable levels of performance become significant. These include:

- Is existing telephone wire adequate?
- Is UTP adequately twisted?
- What about splicing?
- Attenuation, cross-talk, and propagation velocity from tightly coupled pairs become noticeable at 10 Mbps and 100 Mbps, whereas they were not noticeable at 1 Mbps.

By using more suitable encoding techniques, as well as redesigning the transceiver or going over to Category 5 twisted pair, these problems can be overcome to an extent.

In recent times, the LAN switch (intelligent hub) has attracted considerable interest. This device was discussed in Chapter 8. However, as a

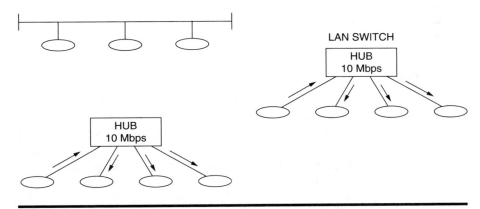

Figure 9.6   Switched Network Topology

quick review, you can consider the switch to represent a multiport bridge that operates based upon the Three Fs — flooding, filtering, and forwarding.

Figure 9.6 depicts the topology of a switched network. The main problem with a hub-based star LAN is that the overall bandwidth of the network is subdivided between the connection pairs.

For example, if a hub is connected to N stations, each connection can avail of at most $N/2$ times the overall bandwidth, as each signal is broadcast to all other ports. With LAN switches, signals are not broadcast. Instead, they are forwarded intelligently to the appropriate destination port. As a result, bandwidth is not shared. Instead, bandwidth can be considered to be dedicated at the overall level to each connection. Assuming the LAN switch has the capacity to keep up with all the attached devices, the overall throughput of the switch is $N/2$ times the bandwidth of the individual connections, because each connection through the switch involves at least two ports.

Note that when upgrading to a switched LAN from a bus LAN, no additional changes are required in hardware or software. The earlier Ethernet MAC protocol continues to hold and there is no change in the media access logic.

9.3   PROS AND CONS OF DIFFERENT TOPOLOGIES

This issue of different topologies can be discussed in terms of a number of LAN characteristics:

- Points of failure
- Intelligence in nodes (controller and stations)
- Recoverability

- Cost of devices
- Performance levels
- Overall goodness

9.3.1 Loop

The main advantage of such LANs is that terminals need to be minimally sophisticated. Controllers use a poll and select mechanisms. Although the actual geometry of such a network is debatable, the main problem would appear to be the bandwidth hogging by greedy or rogue flows, which then starve the other nodes and their connections. Depending on the geometry, a failure can potentially occur in two places in a LAN:

1. At the controller
2. At individual nodes

The first type of failure will incapacitate the entire LAN. The second has an effect that depends on the geometry, but, in general, is easily recovered from. Is it possible to build in sufficient intelligence into the controller?

In this connection, consider the DPMA MAC, as well as the round robin polling scheme discussed later in this chapter. Due to controller failure and other problems, the use of loop LANs has been restricted to niche areas and has generally been excluded from any serious deployment in the industry.

9.3.2 Bus or Tree

Sophisticated MAC protocols are called for in situations where several stations talk simultaneously, which leads to collisions and subsequent errors in transmissions. This is typically a shared-media broadcast situation (unlike rings, which are sequential). As a single point of failure can break up the bus LAN into two subsegments, which can in theory continue to function as individual bus or tree segments. Ethernet cards are available at approximately $20 to $150 and a 16-port 10 Mbps Ethernet hub may be available for as little as $120. In contrast, Token Ring hardware is more expensive, which is possibly why Token Ring networks (and ATM LAN) have gone rather like the dodo bird. In a tree structure, if the head end fails, it renders the network inoperative. Also, the tree may need propagation of interstation frames through a considerable distance, which can introduce a sizable response time delay between stations far away from the head end.

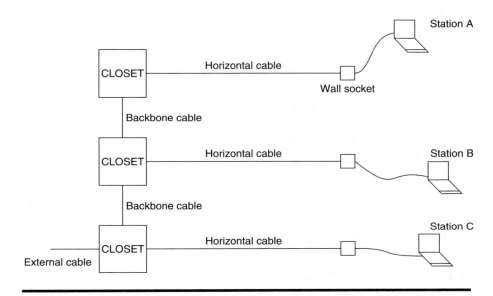

Figure 9.7 An Example of Structured Wiring

9.3.3 Star

In a star topology, the central network controller can be viewed as esentially a PBX switch, as access from one station to another can occur only through the controller. Most commonly this occurs through the use of a LAN switch. The switch supports the simultaneous occurrence of several client/server connections. As will be obvious, switch failure can cause the inoperability of the entire star network. On the more positive side, the star network topology already exists in most buildings with voice-grade lines from offices to the switchboard. So, a LAN that can use in-place wiring is economical and simple to implement.

9.4 STRUCTURED CABLING SYSTEM

Figure 9.7 illustrates an example of the use of structured wiring. The equipment in the telecommunications closet is connected to equipment in other closets via backbone cabling. The closet equipment is connected to individual stations using horizontal cabling.

We introduce key terminology in relation to structured cabling systems, in accordance with the cabling standard ISO 11801. As ever, the principal motivation behind such a standard is basically interoperability of equipment and products from different sources:

- Backbone
- Horizontal cabling
- Cross-connect
- Campus or building distributor
- Telecommunications closet or floor distributor
- Work area
- Telecommunications outlet

These are quickly introduced now. A backbone is a facility between telecommunications closets or floor distribution terminals, their entrance facilities, and equipment rooms. Horizontal cabling comes between the telecommunications closet and the horizontal cross-connect. A cross-connect is a facility allowing the termination of cable elements. Work areas are where users interact with telecommunications equipment. A telecommunication closet is the device in a work area where horizontal cable ends.

9.5 MAC PROTOCOLS

Why do we need MAC protocols? Consider a LAN as a road highway. At each intersection, vehicles are set to enter the road network. If at a certain point near an intersection, Car A enters the highway and a short distance away Car B joins the highway, given that there are no brakes in the internet, a collision will occur, resulting in neither car reaching its destination. Collision control is primarily the essence of LAN MAC protocols.

Before we delve into the details of MAC protocols, we need to first distinguish between two types of devices attached to a LAN:

1. Listener
2. Talker

Listeners are passive devices that only read information meant for them, for example, printers attached to a network. Talkers, in addition to receiving information, send information out as well. Examples are PCs or workstations. A network consists of several talkers and listeners attached.

If two talkers send information within a short time of their times out, the frames sent will interfere and collide with each other, with neither getting to their destination.

9.5.1 Carrier Sense Multiple Access with Collision Detection

CSMA/CD is a listen then send algorithm designed for Ethernet. When a station has a frame to send, it first listens to the network to determine if preexisting traffic is occurring. This typically happens via the bus interface

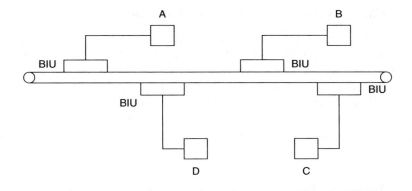

Figure 9.8   CSMA/CD Protocol

unit (BIU), which has an innate capacity to sense the network. This happens in two ways:

1. For a broadband network, the absence of a carrier is interpreted as a free network.
2. For baseband network, the absence of continuous transitions in the middle of each bit functions essentially as a carrier and is interpreted as an absence of digital traffic.

See Figure 9.8 for an example of the CSMA/CD protocol.

Once a frame is lodged into the network, collisions are still possible. For example, considering Figure 9.8, assume Station A sends a frame to Station B. At time T_1, Station A sends, as the channel is free. But at the same time T_1 or near then, Station C also sends to Station D. At time T_1, the channel is sensed free by Station A. But shortly afterward, there is a collision, resulting in the loss of both frames. Collisions occur based on the signal's propagation delay, as well as the distance between competing workstations. When a collision occurs, the MAC protocol needs to recover. This happens, once a collision is sensed, by each competitor backing off randomly or by a predefined timeout.

Today, CSMA/CD is standardized, based on licensed technology from Xerox Corporation. Whereas there is not 100 percent compliance for Xerox's original specification for CSMA/CD, there is a high level of similarity between Xerox and the standardized IEEE 802.3 document. The value of CSMA/CD has been recognized and is the basis of both Fast Ethernet and the more recent Gigabit Ethernet. Networks with bursty bandwidth are best suited for CSMA/CD, as high-volume traffic LANs have more probability of a busy network when a station wishes to transmit, which adversely affects throughput.

9.5.2 Carrier Sense Multiple Access with Collision Avoidance

CSMA/CA is essentially a tweaked version of CSMA/CD. In this scenario, the BIUs attached to the network's talkers estimate when a collision is likely to occur and preclude transmission during those times. The BIUs thus do not need expensive collision-detection capability built in; hence, the cost of hardware to support CSMA/CA is less than that for CSMA/CD. However, the rapid acceptance of Ethernet and the attendant CSMA/CD protocol has rapidly led to the displacement of CSMA/CA as a MAC protocol for LANs.

9.5.3 Token Passing

Token passing is used on a sequential logical topology, whereby if a station wishes to send frames to another station, the frame makes a perambulation of the ring with control information indicating the source and destination addresses, as well as a bit indicating that the token frame has or has not been read and copied by the receiver. The receiving station, when rereading the dispatched frame, notes that the frame has been delivered, absorbs the frame, and generates a free token.

The token passing protocol works for both the Token Bus and the Token Ring and is mentioned for the sake of completeness.

9.5.4 Switched, Connection-Oriented MAC

This method pertains to a switched LAN. A sending station, A, establishes a dedicated path or route to the receiving station, B, via one or more switches. Data from Station A to Station B is sent on that route in a dedicated manner, with sharing occurring only when two such routes are interleaved.

The most common use of a switched connection-oriented protocol is the ATM protocol, which we do not consider in this book for reasons noted earlier.

9.5.5 Demand Priority Media Access

This MAC protocol was invented in conjunction with the 100VG-AnyLAN network architecture. It eliminates the possibility of collisions and subsequent retransmissions. It bears some resemblance to IBM's Loop (Poll and Select) protocol introduced earlier and operates as follows.

Special purpose hubs control all access to the network. Hubs are organized hierarchically, in a cascading fashion, and ports on each hub are assigned a priority, with this mechanism allowing for the differential

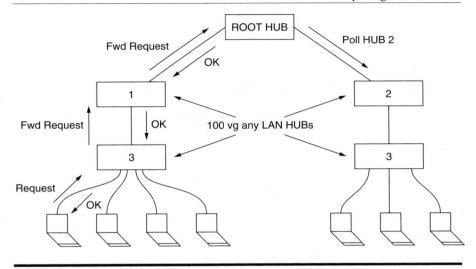

Figure 9.9   The DPMA Setting

management of high-priority traffic, which makes them suitable for multimedia traffic.

This arrangement of hubs and ports is depicted in Figure 9.9.

In Figure 9.9, the following sequence of actions causes the following results:

1. Station requests permission to transmit from local hub.
2. Local hub passes on request to higher layer hub.
3. Higher layer hub passes on request to root hub.
4. Root hub instructs local hub to service the request from station.
5. Root hub polls other higher layer hubs in round robin fashion.
6. High priority ports cannot dominate the network, with lower priority ports sometimes promoted to high priority.
7. Between two stations, a maximum of four hubs are permitted.

9.6   LAN ARCHITECTURE EVOLUTION

Data processing equipment attached to a LAN may be classified in a general way as:

- PCs and workstations — modest data rates
- Server farms — substantial data rates
- Mainframes — high, bulk data transfer

Naturally, it may not be feasible to generate one umbrella LAN that caters to the needs of all the above players in the LAN scenario. As a

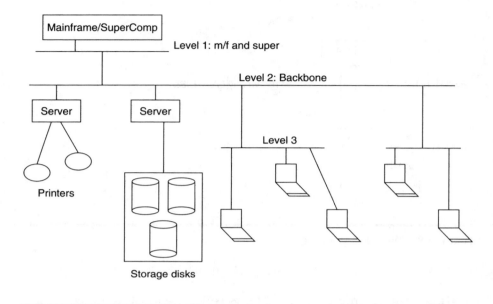

Figure 9.10   An Example of a Leveled LAN Architecture

result, leveled architectures have evolved in relation to the placement of such devices.

Figure 9.10 illustrates an example of a leveled LAN architecture. In this example, we can see the following equipment at different levels:

- Level 1 — supercomputer and mainframe support
- Level 2 — LAN backbone (this connects to server farms as well as to Level 3 clusters)
- Level 3 — clusters of workstations

Level 3 networks are modest in cost and speed. The backbone is faster and more expensive, connecting to shared systems and the Level 3 clusters. In comparison, the connection to mainframe or supercomputers is a special, high-speed link that is more costly than the other connections.

9.6.1   Architectural Design of LANs

Similar to programming, there are two approaches to LAN design:

1. Bottom-up
2. Top-down

9.6.1.1   Bottom-Up

This approach maps well onto the three-level architecture described above. Individual departments or groups within the company understand their needs best and have sufficient budgets to be able to procure a simple Level 3 cluster LAN for their internal needs. Therefore, these usually appear first. In the course of time, several such groups emerge and need to communicate with each other. So there are more clusters and then the decision is made to lash together these clusters. Also, expensive shared peripherals as well as shared servers (server farms) appear commonly to the clusters and a Level 2 backbone LAN is launched to connect all these beasts. Finally, the connection to a mainframe or supercomputer is established.

Because Level 3 networks are designed by local staff, they may be best suited to the department's needs. On the other hand, perhaps if the decision to install LANs were made more centrally, then less overall equipment might be required. Large volume purchases also command more appealing terms from suppliers. Finally, if the Level 3 clusters are from different vendors, there is the ineluctable interoperability issue.

9.6.1.2   Top-Down

The problems associated with the bottom-up approach resulted in many persons favoring a top-down design approach. Under a top-down approach to architecturing the enterprise LAN, a centralized approach is taken along with a total LAN strategy. The advantage is the interoperability of the lower level clusters. The disadvantage is that the end solution may be less than fully responsive to the needs of staff at lower level departments of the company.

9.7   GEOMETRIC LIMITATIONS OF LANS

The need to preserve the sanctity of transmitted messages regarding attenuation, cross-talk, and interference requires that a number of constraints be imposed on the network. For one member of the Ethernet family, namely bus-based coax Ethernet, the specifications and constraints are as follows:

- 50-ohm thick coax cable
- Maximum length per segment = 500 m
- Maximum number of transceivers or nodes per segment = 100
- Minimum separation between nodes = multiples of 2.5 m
- Total number of nodes in the entire network = 1024
- Maximum length of the entire network = 2.8 km
- Speed = 10 Mbps
- MAC protocol = CSMA/CD

These constraints apply verbatim to 10BASE-5.

Moving on, 10BASE-2 also shares these limitations, with the maximum length per segment reduced to 200 m. The speed remains same.

10BROAD-36 is an analog transmission technology and the maximum length per segment is 3.6 km. The other two members of the initial Ethernet family, 10BASE-5 and 10BASE-T, run over twisted pair wire and share a star topology, rather than a bus topology, for the three earlier introduced Ethernets. Twisted-wire-based networks are less expensive than bus-based coax and are inherently more reliable.

Consider two segments, one a bus-based coax segment and the other a star-based transaction procession (TP) segment. A failure on the bus can, depending on where it occurs, incapacitate the entire bus. On the other hand, the failure of a cable on a star-based network only affects the station cabled to the hub. However, a hub failure renders the entire network inoperative but, being sheltered inside telecommunications closets, such hubs are protected from damage and repair.

Clearly, for similar configurations, achieving a 100 Mbps throughput on a LAN seems quite an attractive proposition. These Fast Ethernets, generically called 100BASE-T, have three possible incarnations — 100BASE-TX, 100BASE-FX, and 100BASE-T4. These are discussed in detail in Chapter 10. Each supports the use of the generic CSMA/CD MAC protocol.

But where does the extra bandwidth come from?

- Transmit on 3 pairs versus 1 pair ×3.00
- 8B6T coding instead of Manchester ×2.65
- 20 to 25 MHz clock increase ×1.25
- Total throughput increase ×10.00

9.8 FIBER CHANNEL TOPOLOGIES

Fiber channel is a high-speed LAN solution that can connect systems spread over as far as 10 km at speeds up to 3.2 Gbps in topologies, as depicted in Figure 9.11. These topologies include point-to-point, loop or hub, and switched. Frame sizes vary from 0 to 2 Kb and the solution uses optical fiber cable (OFC) and copper.

Fiber channel is intended to provide full-duplex links (fibers per link). The entire system is designed to be delay insensitive and broadly available (standard components) supporting multiple price-to-performance levels ranging from small systems to managing supercomputers.

We discuss fiber channel in detail in Chapter 10, looking here at only an overview with the associated topological considerations involved.

The fiber channel network supports five layers:

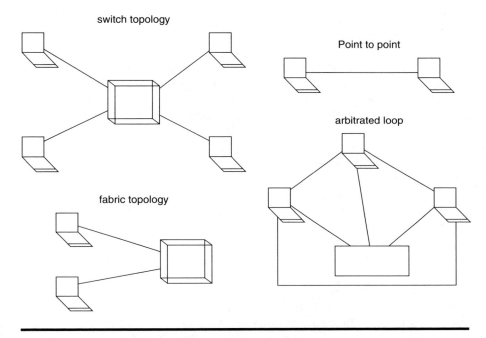

Figure 9.11   Fiber Channel Topologies

- FC-0 — physical interface
- FC-1 — transmission layer (byte synchronization and encoding)
- FC-2 — signaling protocol (actual transport mechanism)
- FC-3 — common services
  - Striping (parallel node-ports)
  - Hunting (similar to hunting lines in telephone company)
  - Multicast
- FC-4 — mapping to upper layer protocols:
  - Multimedia
  - Channels
  - Internet protocol interface (IPI)
  - High-performance parallel interface (HIPPI)
  - Small computer systems interface (SCSI)

Fiber channel is more like a traditional packet switched network in contrast to shared media access LANs. There is no need for a MAC protocol and the solution scales beautifully in terms of number of nodes, data rates, and distance covered, by simply adding switches.

Key terms:

- Fabric
- N-port

- F-port
- NL-port

The fabric is the entity that interconnects various attached nodes and manages the routing of frames. Each node has one or more ports called N-ports. Similarly, each fabric (switching) element has one or more F-ports. All routing of frames between N-ports is done by the fabric.

Fiber channel media supported include:

- Single mode fiber
- 50 μm multimode fiber
- 62.5 μm multimode fiber
- Video coax cable
- Miniature coax cable
- Shielded twisted pair (STP)
- Twinax

Each choice of media has an associated set of data rates, as well as maximum point-to-point link coverage. As noted, data rates range from 100 Mbps to 3.2 Gbps and distances from 100 m to 10 km.

There are four possible topologies as indicated in Figure 9.11:

- Fabric and switch topologies
- Point-to-point
- Arbitrated loop
  - Without hubs
  - With hubs

Routing in a fabric topology is transparent to the nodes. Each port in the system is uniquely addressed. The fabric is intelligent enough to switch incoming frames. Also the fabric is scalable, with added ports leading to additional switches and increased fabric capacity. The fabric is largely protocol independent and delay insensitive. Minimal demands are placed on nodes.

In a point-to-point topology, there is no routing as individual nodes are directly linked.

Finally, in the arbitrated loop topology, we have a simple, low-price method to connect up to 126 nodes in a loop that functions in a similar manner to a Token Ring situation, via a token passing mechanism.

Topologies, media, and data rates are selected by the LAN designer to optimize the service provided to a given design situation.

10

A TUTORIAL ON THE ETHERNET FAMILY OF LOCAL AREA NETWORKS

10.1 INTRODUCTION

In this chapter, we turn our attention to the several members of the Ethernet family, including:

- 10 Mbps Ethernet
- Fast Ethernet
- Gigabit Ethernet
- 10 Gigabit Ethernet

Each member of the family has its immediate relatives, for instance, within 10 Mbps Ethernet, we find 10BASE-T, 10BASE-2, 10BASE-5, 10BROAD-36, and 10BASE-F. Within 100 Mbps Fast Ethernet, we find 100BASE-TX, 100 BASE-FX, and 100BASE-T4. Also, in 1000 Mbps Ethernet, better known as Gigabit Ethernet, there are 1000BASE-LX, 1000BASE-SX, 1000BASE-CX, and 1000BASE-T. However, in 10 Gbps operations, better known as 10-Gigabit Ethernet, only one standardized LAN is defined.

Recall that a LAN solution consists of a transmission medium, MAC protocol, and encoding mechanism and operates using a predefined topology. The transmission medium is concerned with the properties of the physical carrier that bears the signals from source to destination. The MAC protocol governs the method by which signals access the medium. The encoding mechanism defines how data and control codes are coded. Because all of the LAN technologies we discuss are baseband (digital transmission of digital data), we will be concerned with only digital encoding. Different signal elements are used to represent binary 1 and binary 0. A number of encoding schemes are discussed later in this chapter.

In previous chapters, we looked at all these components, paying particular attention to topology and MAC protocols in Chapter 9. The other two components of LANs — transmission media and encoding schemes — were only mentioned there in passing, primarily for the sake of completeness. We are able to afford a more detailed treatment of these aspects of LANs in this chapter.

10.2 TRANSMISSION MEDIA

Let us first turn our attention to the transmission media. There are primarily five possibilities:

1. UTP
2. STP
3. Coaxial cable
4. OFC
5. Unguided (wireless)

The last medium is addressed in Chapter 13 on Wireless LANs.

Each medium has its benefits and limitations. Whatever the medium, the following key considerations and characteristics must be considered:

■ Bandwidth (note that the higher the bandwidth, the higher the data rate)
■ Transmission impairments
■ Mutual interference
■ Cost
■ Ease of installation
■ Geographic scope supported
■ Maximum speed of communication supported

During the ensuing discussion, we will have an opportunity to make observations about each of these criteria.

10.2.1 Twisted Pair

Twisted pair comes in two varieties:

1. UTP
2. STP

UTP is the least expensive type of twisted pair wiring. Office buildings come prewired with a lot of excess 100 ohm voice-grade UTP wires. Because twisted pair easily bends around corners and is commonly located near office desks, this medium is both readily available and easy to install. However, UTP is susceptible to considerable interference from external fields and picks up a lot of noise as well.

To improve the performance of UTP, STP was developed. This medium shields each pair of twisted wire using a metallic sheath to reduce interference. This improves performance, but is more expensive than UTP. Also, it is not as easy as UTP to bend around corners.

The initial interest in voice-grade UTP media was to provide services at 1 to 16 Mbps and was adequate and suitable for applications extant at that time.

However, since that time, users have migrated to higher bandwidth applications and their requirements of the LAN have moved up considerably. Typically, 100 Mbps Fast Ethernet and 1000 Mbps Gigabit Ethernet have become necessary, so ways had to be found to upgrade LAN performance to these levels. To deal with these new requirements, three types of UTP cable were initially standardized jointly by the Electronics Industry Association (EIA) and the Telecommunications Industry Association (TIA) within their joint EIA/TIA-568 standard. The three types of cable are:

1. Cat(egory) 3
2. Cat(egory) 4
3. Cat(egory) 5

Along with the cables and the associated hardware, the speeds supported by these media include:

- Cat 3 — up to 16 MHz
- Cat 4 — up to 20 MHz
- Cat 5 — up to 100 MHz

Note that the ability of these categories of UTP to support different types of LAN transmission will depend on the signaling method used by different LANs. For instance, consider a 4B/6B encoding. In this case, a Cat 5 cable will support a transmission rate of $(6/4) \times 100 = 150$ Mbps.

10.2.2 Coaxial Cable

Coaxial cable consists of a pair of conductors, but is constructed differently to allow the operation of a broader spectrum of frequencies.

Within a coaxial cable is a concentric pair of conductors, with the inner conductor providing a single wire surrounded by a dielectric insulator. The insulator is in turn covered by the second hollow conductor ring. The second ring conductor is protected by a jacket, which forms a shield. Because of the shielding, coaxial cable is less susceptible to noise and interference than twisted pair. Greater distances are possible as well, as is the support for more attached stations. For example, coaxial cable supports hundreds of Mbps over transmission distances of 1 km. Although coaxial cable is more expensive than STP, it provides greater capacity. However, coaxial cable is less flexible than twisted pair in terms of bending around at the point of connection.

10.2.2.1 Coaxial Adapters

This device connects an existing (thin) coaxial cable bus network (up to 29 stations) with a wire hub. The maximum network span is 100 m. The purpose of the adapter is to connect the BNC T-connector used with 10BASE-2 to the UTP cable whose other end is attached to the wire hub.

Essentially the coaxial adapter is a two-port repeater connecting one 10BASE-T port and one thin coax BNC port (10BASE-2). Through its use, a thin coax cable (200 m length and 29 stations) can be integrated into a 10BASE-T network without any modification to the existing coaxial network infrastructure.

The 5-4-3 rule applies to each member of the 10 Mbps Ethernet family. As a refresher, the rule indicates that data frames can traverse a maximum of three (3) populated segments, four (4) repeater hops, and five (5) total segments. Any segment with a workstation represents a populated segment.

10.2.3 Optical Fiber Cable

OFCs are made of three possible substances, in order of decreasing cost and performance:

1. Ultra pure fused silica
2. Multicomponent glass
3. Plastic fiber

An OFC consists of three concentric rings:

1. Core
2. Cladding
3. Jacket

The core is the central section composed of many fibers. Each fiber is surrounded by a cladding consisting of either a glass or plastic coating.

The outermost ring is composed of plastic-like materials to protect the cladded fibers.

OFC is being used for long-haul telecommunications, as well as in LAN environments. The ongoing improvements in technology and the dropping cost to manufacture are making OFC more and more popular as an alternative medium for LANs. The key advantages for the use of optical fiber include a significantly higher bandwidth, which permits a much higher data rate than obtainable on copper-based media, its micro compact size, and immunity to electronic interference from external sources.

OFC systems cover both the infrared and visible spectra. The different types of OFC technologies include step index multimode, single mode, and graded index multimode. Once a light pulse enters an OFC, it behaves according to the physical properties of the core and cladding.

In multimode step index fiber, the light bounces off the cladding at different angles, continuing down the core, whereas others are absorbed by the cladding. This type of OFC supports data rates up to approximately 200 Mbps for distances up to 1 km.

By gradually decreasing the refractive index of the core, reflected rays are focussed along the core, more efficiently yielding data transmission rates of up to 3 Gbps over many kilometers. This type of optical fiber is referred to as graded index multimode.

The last type of optical fiber focuses rays even farther, so that only one wavelength can pass through the core at a time. This type of fiber is referred to as single mode and is the most expensive, as well as best performing. Lasers can be coupled to single mode fiber, permitting extremely high data rates at long distances.

10.2.3.1   Fiber Optic Technology

A significant enhancement to the 10BASE-T Ethernet technology is Fiber Optic Repeater Link (FORL). Transmission of data occurs across dual fiber cable (1 for transmission, 1 for reception). OFC technology enables the support of multiple Ethernet segments at distances up to 2 km. Using a fiber transceiver, one can connect remote stations, connect a wire hub and a fiber hub, and support multiple stations.

When an optical transceiver is used on a wire hub, one can connect to dual fiber cable. Basic OFC devices are briefly described here mainly for the sake of completeness and continuity of the discussion.

10.2.3.1.1   Optical Transceiver

This device consists of electronics and circuitry that translate ON and OFF indicators into the presence and absence of light signals, which in turn are mapped to an encoding scheme.

10.2.3.1.2 Fiber Hubs

Fiber hubs consist of many FORL ports, one AUI port, and one or more 10BASE-T ports. The FORL ports link fiber hubs to fiber adapters or fiber adapters to fiber NAUs in PCs.

10.2.3.1.3 Fiber Adapters

A fiber adapter is a media conversion device, translating between coaxial and optical fiber. The fiber adapter extends the transmission distance between a wire hub and a station from 100 m to 2 km, with an adapter required at each end of a fiber link, unless a station is directly connected to a fiber hub. When attached to a fiber hub, the distance separation is 2 km. When attached to a wire hub, the maximum transmission distance is reduced to 15 m. When attached to a PC's NAU, the separation is again 2 km.

10.3 AN EXCURSION INTO THE ETHERNET FAMILY

In this section, we look at the various members of the Ethernet family, including the 10 Mbps LAN series of standards, 100 Mbps Fast Ethernet, 1000 Mbps Gigabit Ethernet, and the 10 Gigabit network. The common IEEE notation for these network solutions is:

```
< Data Rate Mbps> < Encoding method> < Maximum
segment length in hundreds of meters>
```

Therefore, using this notation, we can illustrate that a 10BASE-2 network represents a 10 Mbps baseband network with a maximum segment length of 200 m. In actuality, the maximum length of a 10BASE-2 network is 185 m, but the IEEE nomenclature refers to the standard as a 200 m maximum length.

10.3.1 10 Mbps LANs

There are four 10 Mbps Ethernet LANs standardized by the IEEE:

1. 10BASE-5
2. 10BASE-2
3. 10BASE-T
4. 10BROAD-36
5. 10BASE-F

10.3.1.1   10BASE-5

The media cable used for this network is 50 ohm coaxial cable, which means less interference (low noise) and less reflections. The data rate supported is 10 Mbps. The maximum segment length is 500 m. 10BASE-5 is extensible using repeaters, with a maximum of 4 repeaters between any 2 stations on the LAN based on the previously mentioned 5-4-3 rule. The maximum LAN length itself is 2.5 km. The topology is a bus structure resulting in stations contending for access to the bus. The cable diameter is 10 mm and, at most, 100 nodes are permitted per segment. The maximum number of nodes per network is 1024 and the maximum node spacing is 1000 m. The cable and wire type is more formerly referred to as RG-50/58, with N-type connectors.

10.3.1.2   10BASE-2

A 10BASE-2 network represents a less expensive and less capable LAN in comparison to 10BASE-5. In a 10BASE-2 network, the electronics are attached to the station without an AUI. Although this network uses a 50 ohm coaxial bus topology cable with the same data rate of 10 Mbps as a 10BASE-5 network, there are significant differences between the two. First, the 10BASE-2 cable diameter is 5 mm, which means it is significantly thinner than the cable used in a 10BASE-5 network. Because of this, the cable used in a 10BASE-2 network is sometimes referred to as thinnet or cheapnet cable. Because of the thinner cable, the maximum segment length is 185 m, with up to 15 m allowed for the tap, resulting in a segment length of 200 m. The 10BASE-2 network permits a maximum network span of 1000 m, with 30 nodes per segment allowed, as well as 1024 maximum nodes per network. Minimum and maximum node spacing on a 10BASE-2 network are 0.5 m and 200 m, respectively. The maximum number of segments is three.

Note that 10BASE-5 and 10BASE-2 can be interconnected. However, because 10BASE-2 is less noise resistant, unexpected problems can occur when these two types of networks are bridged. The type of cable used for a 10BASE-2 network is more formally referred to as RG-6/6a/22, with BNC connectors used to attach to the cable.

10.3.1.3   10BASE-T

The 10BASE-T LAN is a twisted wire hub centric network that permits the use of the prewired installed base of UTP cable in most organizations. Because the 10BASE-T network is hub-based, the topology can be viewed as a star.

The data rate of a 10BASE-T network is again 10 Mbps, and the encoding is Manchester. The maximum link length is 100 m. 10BASE-T is interoperable with 10BASE-2 or 10BASE-5 networks. The maximum number of coaxial cable segments in a path between between stations is three. The maximum number of repeaters is four and the maximum number of segments is five. Thus, this follows the 5-4-3 rule previously mentioned in this chapter.

10.3.1.4   10BROAD-36

As previously noted, 10BROAD-36 uses radio frequency modems and does not encode data digitally. Because this technology is largely superseded and does not use digital encoding it will not be discussed further.

10.3.1.5   10BASE-F

The 10 Mbps version of Ethernet that operates over optical fiber is referred to as 10BASE-F. This network provides many benefits associated with the use of optical fiber and can be implemented in three versions:

1. 10BASE-FP — this is a star situation using stations and repeaters up to 1 km per link.
2. 10BASE-FL — point-to-point connections of either stations or repeaters up to 2 km.
3. 10BASE-FB — defines a point-to-point link that can be used to connect repeaters at up to 2 km.

The media for all three versions of 10BASE-F is an optical fiber pair, with encoding occurring using Manchester encoding. 10BASE-FP supports 33 stations per star.

10.3.2   Fast Ethernet (100 Mbps)

Fast Ethernet provides a low-cost Ethernet networking capability at a data rate of 100 Mbps. This version of Ethernet uses the same frame format as Ethernet 802.3 LANs, as well as the same MAC protocol. The topology is a star, because it is based on the use of hubs.

Figure 10.1 illustrates what we refer to as the Fast Ethernet tree and indicates the different versions of this networking technology, as well as the different media the versions support. As indicated in Figure 10.1, Fast Ethernet comes in three basic versions:

1. 100BASE-TX — the topology is a star and the encoding method is MLT-3. The media used are either two pairs STP or two pairs Cat 5 UTP. The maximum segment length (link) is 100 m. The total network span is 2500 m.

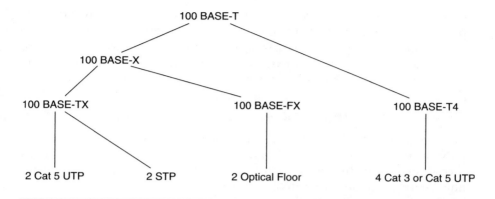

Figure 10.1   The Fast Ethernet Tree

2. 100BASE-FX — topology is a star, with data encoding being either 4B5B or NRZ-I and uses two optical fibers, one for each direction.
3. 100BASE-T4 — topology is a star, with the encoding being either 8B6T or NRZ. It is the most popular version of Fast Ethernet as it supports the use of either four Cat 3 or Cat 5 UTP cables.

Each version of Fast Ethernet uses the same MAC protocol (CSMA/CD) and framing as their 10 Mbps cousins. The only difference between the two families concerns the envelopes of framing at 100 Mbps, with Fast Ethernet using special codes referred to as starting and ending delimiters. Because those delimiters are ignored by network adapters, the resulting framing is considered to be the same as on a 10 Mbps Ethernet network.

One key difference between 100 Mbps Fast Ethernet and its 10 Mbps cousins concerns the 5-4-3 rule. Specifically, the 5-4-3 rule does not apply to Fast Ethernet. Cable distance is restricted to a maximum of 100 m and without optical fiber technology, the maximum distance between nodes is 205 m. If two Fast Ethernets are connected, the maximum distance between the networks is restricted to a maximum of 5 m. Thus, Fast Ethernet networks cannot be cascaded.

There are two key aspects to selecting the architecture of a Fast Ethernet network:

1. Backbone operation
2. Switch segmentation

10.3.2.1   Backbone Operation

In this context, a 100 Mbps Fast Ethernet hub is used with autoconfiguring 10/100 Mbps ports to connect two or more 10BASE-T networks. Unfortunately, this scenario results in one big collision domain, without any performance improvement for the users on the horizontal axis.

If some users are relocated from the 10BASE-T hub to the backbone hub, those users can avail of the 100 Mbps rates in the backbone. This occurs specifically when servers are placed on the backbone.

10.3.2.2 Switch Segmentation

For true performance improvement, replace the 100BASE-T hub with a LAN switch, which functions broadly, as described in Chapter 8, by providing N/2 connections each at 100 Mbps, where *N* represents the number of ports on the switch.

You can use a switch to provide connectivity with both departmental servers and individual stations. The former are attached directly to the switch and are serviced at 100 Mbps, whereas the latter are placed on 10 Mbps LAN segments. This design technique provides additional bandwidth as appropriate to departmental servers, whereas the total bandwidth is shared on the 10 Mbps LAN segments.

10.3.3 Gigabit Ethernet (1000 Mbps)

Gigabit Ethernet is a LAN technology that allows for transmission at data rates of 1000 Mbps. Figure 10.2 illustrates the architecture of a Gigabit Ethernet LAN, with a 1 Gbps LAN switch servicing three 100 Mbps Fast Ethernet hubs, while a server or perhaps a server farm is supported directly at a 1 Gbps data rate.

Gigabit Ethernet uses the same CSMA/CD MAC protocols as Fast Ethernet and Ethernet, which makes all three interoperable.

There are four physical layers supported by the Gigabit Ethernet series of standards. Those standards and their physical media used are summarized below:

1. 1000BASE-LX — 1300 nm laser on single/multimode fiber
2. 1000BASE-SX — 850 nm laser on multimode fiber
3. 1000BASE-CX — short-haul copper twinax STP cable
4. 1000BASE-T — long-haul copper UTP

Data transferred to a specific type of Gigabit media are encoded using either 8B/10B or 4D PAM5. This is important to know from a design standpoint as it has a bearing on the data rates.

The basic Gigabit Ethernet architecture is depicted in Figure 10.3. Encoded data are transmitted to the MAC layer, with the exception of the UTP Gigabit Ethernet where a special GMII (Gigabit Media Independent

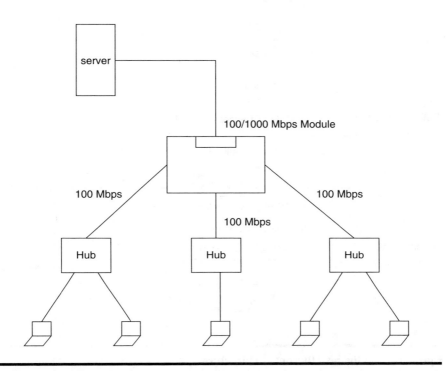

Figure 10.2   Server/Switch Connection

Interface) is defined to connect the physical and MAC layers. The GMII is an interface that provides 1 byte parallel receive and transmit as a chip-to-chip synchronous interface. The GMII is divided into three sublayers:

1. Physical coding sublayer (PCS) — provides a uniform interface to the reconciliation layer for all physical media. 8B/10B encoding is used. Carrier sense and collision detection are functions of the PCS. It also supports the autonegotiation process for NICs.
2. Physical medium attachment (PMA) sublayer — provides a medium-independent means for the PCS to support various serial bit-oriented physical media.
3. Physical medium dependent (PMD) sublayer — maps the physical medium to the PCS. It defines the physical layer signaling for various media.

Unlike when using a 1 Gbps LAN switch, which does not require a separate MAC protocol, because of the available full-duplex operation, when using Gigabit Ethernet on LAN interconnecting devices, an enhancement is required to the basic CSMA/CD scheme. This occurs as:

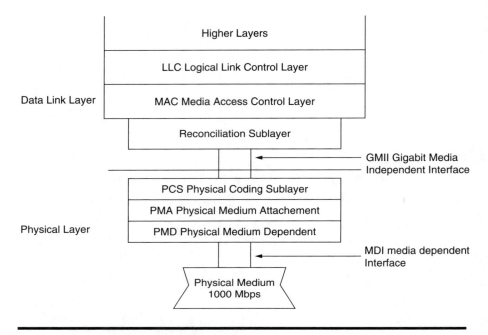

Figure 10.3 Gigabit Ethernet Architecture

■ Carrier extension
■ Packet bursting

Carrier extension represents a way of maintaining the IEEE 802.3 minimum and maximum frame sizes plus decent cabling distances.

Figure 10.4 illustrates the Gigabit Ethernet frame format to include carrier extension. The carrier extension are nondata extension symbols included as a padding within the collision window to ensure the minimum length frame is 512 bytes. The entire padded frame is considered for collision detection, whereas only the original data without extension are used for error check frame check sequence (FCS) considerations.

Note that carrier extension wastes bandwidth with too many extension symbols per actual meaningful information sent. Small packets may use up to 448 pad bytes.

But with packet bursting, a burst of packets are sent. The first of these are padded as before. Subsequent packets are sent back-to-back without extensions, maintaining the interpacket gap and the burst time. This procedure substantially improves throughput. Figure 10.5 shows how packet bursting works.

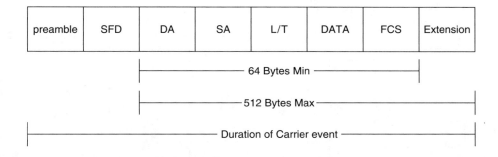

Figure 10.4   Gigabit Ethernet with Carrier Extension
Legend: SFD = Start of frame delimiter; DA = Destination address; SA = Source address; FCS = Frame check sequence.

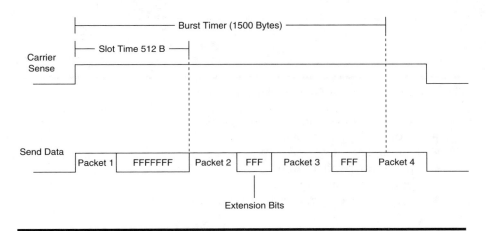

Figure 10.5   Gigabit Ethernet with Packet Bursting

10.3.4   10 Gigabit Ethernet

As the demand for high-speed networks continued to grow, a need for faster Ethernet technology became apparent. In early 1999, the IEEE 802.3 committee chartered the High Speed Study Group (HSSG) to standardize what has come to be known as 10 Gigabit Ethernet. Some of the objectives of the HSSG included:

- Support for 10 G Ethernet at 2 to 3 times the cost of Gigabyte Ethernet
- Maintain earlier frame formats
- Meet IEEE 802.3 functional requirements

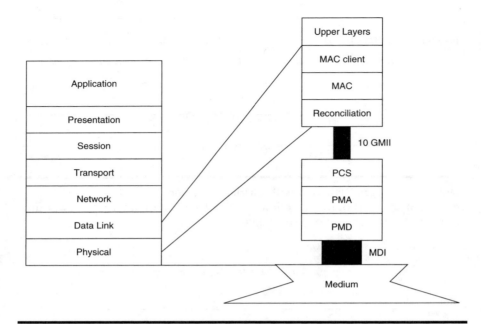

Figure 10.6   10 Gigabit Ethernet Architecture

- Provide compatibility with IEEE 802.3 flows
- Media independent interface
- Full-duplex only
- Speed independent MAC
- Support for star LAN topologies
- Support for existing and new cabling infrastructure

Some of the benefits of 10 Gigabit Ethernet include a low-cost solution for enhanced bandwidth and faster switching. Because there is no need for fragmentation, reassembly, or address translations and switches are faster than routers, using 10 Gigabit Ethernet as a backbone technology provides a mechanism to remove bottlenecks with a scaleable upgrade path.

Figure 10.6 illustrates the architecture of 10 Gigabit Ethernet. Similar to other versions of Ethernet, several layers are defined, some specific to the use of different types of media. Let us examine some of those layers:

- MAC layer — provides a logical connection between the MAC clients of itself and its peer station. Functions include initialize, control, and managing connections.
- Reconciliation sublayer — acts as a command translator. It maps MAC layer terms and commands into the electric formats for the physical layer.

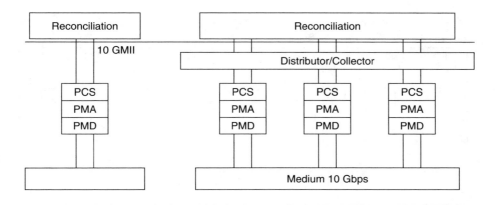

Figure 10.7 10 Gigabit Ethernet Serial and Parallel Implementations

- 10 Gigabit media independent interface (10GMII) — functions as the standard interface between the MAC layer and the physical layer.
- PCS sublayer — codes and encodes data to and from the MAC sublayer. No standard encoding scheme is defined for this layer.
- PMA sublayer — serializes code groups into bit stream suitable for serial bit-oriented physical devices and vice versa.
- PMD sublayer — responsible for signal transmission. Amplification, modulation, and wave shaping functions are performed by this sublayer. Different PMD devices support different media.
- Medium dependent interface (MDI) sublayer — references a connector. The sublayer defines different connector types that attach to different media.
- Physical layer architecture — there are two structures for the physical layer implementation of the 10 GB Ethernet:
  - Serial implementation
  - Parallel implementation

  The former uses one high-speed (10 Gbps) PCS/PMA/PMD circuit block, whereas the latter uses multiple blocks at lower speed. Figure 10.7 depicts these two situations.

Currently, the preferred media adopted for 10 Gigabit Ethernet is optical fiber. Because of its high data rate, it is doubtful that a copper version of the technology can be developed.

10.4 LAN ETHERNET DESIGN

The first step in the Ethernet LAN design process is to select a concentrating device, such as a wire hub, bridge, LAN switch, or router. All can be at

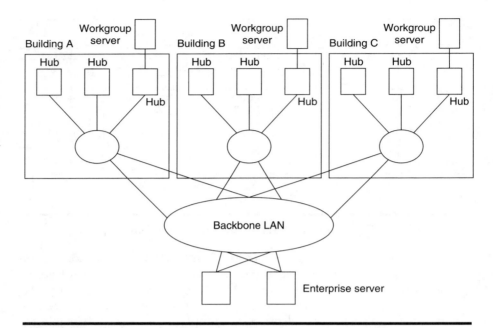

Figure 10.8 Traditional Hub and Router Campus Network

10 Mbps to 1 Gbps speed. Then select a transmission medium (and encoding scheme), all appropriately at 10 to 1000 Mbps speed.

Let us first consider a wire hub and router model, as illustrated in Figure 10.8 where an I/O port on the wire hub enables an extended 10BASE-T network, basically a star topology (extended).

What are the advantages of 10BASE-T? In comparison to coaxial cable, it is less expensive and more flexible, there is an extensive installed base of extra wiring, and being point-to-point, any breakage impacts only one user.

When using 10BASE-T, connectivity to other types of IEEE 802.3 LANs, such as 10BASE-5 and 10BASE-2, occurs via AUI on the hub. There is one AUI port per wire hub, which is used to connect to other IEEE 802.3 networks. Figure 10.9 depicts the interconnectivity between 10BASE-T and 10BASE-5 networks.

The next step in designing LANs is depicted in Figure 10.10, wherein Layer 2 switching is used in the core, distribution, and access layers. There are four workgroups attached to the access layer switches. Router X connects to all four virtual LANs (VLANs). Layer 3 switching and services are concentrated at Router X as well. Enterprise servers are connected logically to Router X. Router X is typically referred to as a "Router on a Stick," serving many VLAN connections.

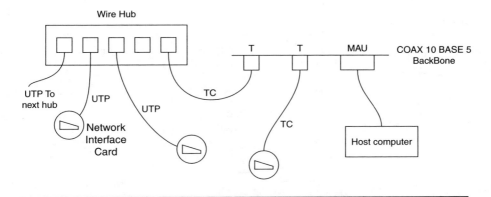

Figure 10.9 Interconnecting 10BASE-T and 10BASE-5 Networks
Legend: MAU = Media attachment unit; UTP = Unshielded twisted pair; TC = Transceiver cable; T = Transceiver

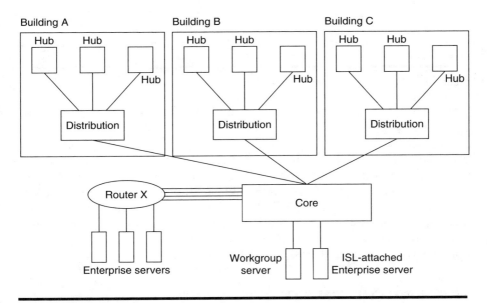

Figure 10.10 Campuswide VLAN Design

10.4.1 Campuswide VLANs with Multilayer Switching

This type of networking structure makes it possible for configured stations to relocate to a different floor or even a different building, e.g., a mobile user plugs a laptop into a different LAN port in another building. Such a

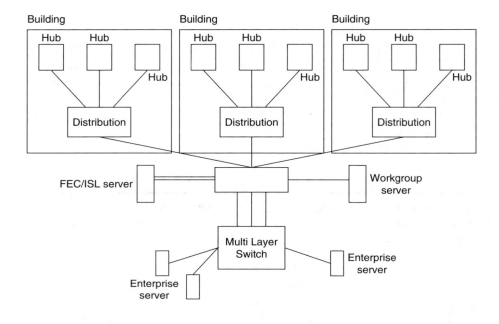

Figure 10.11 Multilayer Switching

situation is typically handled by the use of a VLAN Trunking Protocol (VTP) and is illustrated in Figure 10.11.

We can now see how a computer on a coaxial-cable-based network connects to a wire hub. A transceiver interfaces the transceiver cable much like a host station is connected to a coaxial network. In this situation, the NIC is connected via a transceiver cable to the transceiver on the LAN. However when a coaxial cable NIC is attached to a wire hub, an AUI adapter interfaces the NIC with a UTP that connects to the wire hub. Figure 10.12 illustrates the use of an AUI adaptor so that a NIC can be connected to a wire hub port using a UTP cable.

10.5 SWITCHES REVISITED

Switches are a fundamental aspect of most networks. They allow source and target nodes to communicate over a network at the same rate without slowing each other down by sharing multiple simultaneous connections. Switches have been discussed in detail in other chapters and are revisited briefly here for the sake of continuity.

The following problems were observed with a hub-based network configuration.

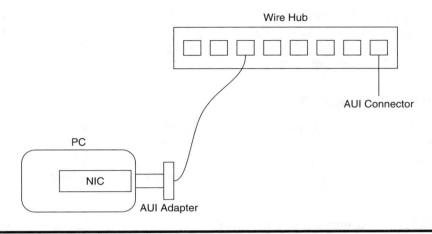

Figure 10.12   Connecting a Coaxial Cable NIC to a Wire Hub

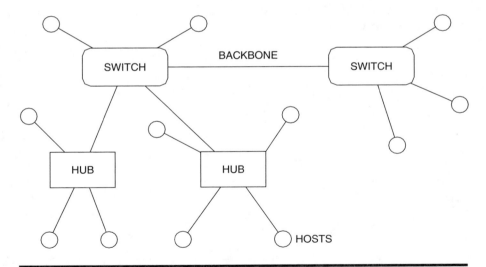

Figure 10.13   Hubs and Switches

10.5.1   Scalability, Latency, Global Effect of Failures and Collisions

By adopting the schematic configuration in Figure 10.13, it becomes possible to alleviate some of the problems associated with a hub-based network configuration.

Switches alleviate many hub-based problems by dedicating bandwidth to individual unicast communication paths or connections. So, if there are N ports in the switch, with each connection at 10 Mbps, the switch itself delivers N/2 × 10 Mbps to the configuration.

10.5.2 Encoding Schemes

In concluding this chapter, we will briefly review LAN encoding schemes as they define how it becomes possible to achieve a given data rate on a particular media that supports a given signaling rate. Various common LAN encoding schemes are described below in order of increasing complexity.

10.5.2.1 Nonreturn to Zero Level

NRZ-L is perhaps the simplest encoding method (see Figure 10.14). Under this encoding scheme, a high voltage is used to indicate the presence of a one bit, and the absence of voltage is used to indicate the presence of a zero bit. Because two or more successive set or nonset bit positions need clocking to differentiate the different bits, it would be expensive to use this coding on a LAN.

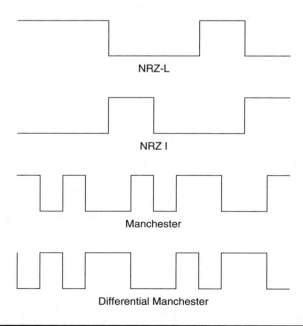

NRZ-L

NRZ I

Manchester

Differential Manchester

Figure 10.14 Some Basic Encoding Schemes

10.5.2.2 Nonreturn to Zero Invert on 1s

Under an NRZ-I encoding scheme, one maintains a constant voltage pulse per bit time. Data is encoded as the presence or absence of a transition at the beginning of a bit time. A transition at bit start signifies a binary 1. No transition is a binary 0. This is a case of differential coding. Its main benefit is that it may be more reliable to detect a transition in the presence of noise than to compare voltage threshold values. Another consideration is that it sometimes becomes easy to lose polarity of the signal. Because there are problems with NRZ, in particular the loss of time when one bit ends and another one begins, which leads to a drift in timing and ensuing corruption of the signal, other coding schemes were introduced.

10.5.2.3 Manchester

In Manchester coding there is a transition at the middle of each bit period. A high-to-low transition is a binary 0, whereas a low-to-high transition is coded as a binary 1. Note that this coding method provides a self-clocking function as individual bits from pairs of set or nonset bits can be distinguished.

10.5.2.4 Differential Manchester

Under Differential Manchester encoding, the mid bit transition does only clocking. The presence of a transition at the beginning of a bit period means a binary 0. No transition at the beginning of a bit period is a binary 1.

Both Manchester and Differential Manchester coding techniques are popular for LANs. They are sometimes called biphase, because there may be as many as two transitions per bit time. So, the maximum modulation rate is twice that of NRZ, which in turn means more bandwidth.

The advantages of Manchester and Differential Manchester coding include synchronization (based on transitions), no DC component, and a built-in error detection capability, because noise needs to invert the bit before and after, which is unlikely.

10.5.2.5 4B/5B-NRZ-I

Under 4B/5B encoding, encoding is done four bits at a time. Every four bits translates into five code bits. The efficiency is thus 80 percent. For further synchronization, each code bit is treated as a binary value and further encoded with NRZ-I. This scheme lends itself to optical fiber transmission.

10.5.2.6 MLT-3

Under MLT-3, three levels are used for encoding a binary 1:

1. A positive (+ve) voltage
2. No voltage
3. A negative (-ve) voltage

The steps involved are:

1. If the next input bit is 0, the next output value persists.
2. If the next input is 1, there will be a transition, as follows:
   a. Preceding value is $+/-$ V → next output value is 0.
   b. Preceding value is 0 → next output is nonzero and opposite in sign to the last nonzero output.

Note that this in turn implies that if the signaling rate is one third of the operating rate, a baud rate of 33.33 MHz will support a LAN at 100 MHz.

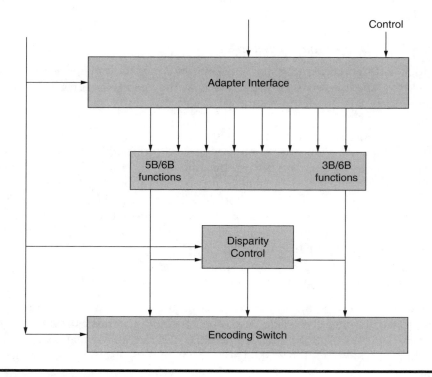

Figure 10.15   8B/10B Encoding

10.5.2.7   8B/10B

8B/10B encoding is popularly used in fiber channel and Gigabit Ethernet. Under this encoding technique, each eight bits of data translates into ten bits of output. 8B/10B was developed and patented by IBM; it is more powerful than 4B/5B in terms of transmission features and error detection.

Figure 10.15 illustrates an example of the generic mB/nB encoding, mapping m source bits into n output bits.

Note the use of a functionality called disparity control. This basically addresses the excess of 0s over 1s or vice versa. An excess in either case is called a disparity. If there is one, the disparity control block complements the 10-byte block to redress the problem.

11

ETHERNET PERFORMANCE CHARACTERISTICS

11.1 INTRODUCTION

In this chapter, we turn our attention to Ethernet LAN performance, using the general concepts of Ethernet as introduced in Chapter 10. First, we will look at the issue of frame sizes and the length of the information field and the overhead of a frame. Therefore, we first deal with, in detail, the composition of a LAN Ethernet frame. Can the length of LAN frames or their information carrying capability be adjusted to achieve enhancement in performance? Likewise, the effect of frame length on bridge and router operations is investigated.

If we have an up and running LAN and wish to expand or enhance it, we can monitor current LAN traffic to predict the effect of an expansion or the enhancement on a similar planned network. But, when a brand new network is being put in place, we lack a prior baseline. In this situation, we need a theoretical framework to estimate network traffic. That framework occurs through the use of a LAN traffic estimation technique. We will explore the use of this technique to predict future network growth and the effect of such expansion on the future planned network, as well as the segmentation of a LAN to improve network performance.

11.2 FRAME OPERATIONS

The key to understanding Ethernet LAN performance is to first appreciate how an Ethernet frame is constructed. Within a frame, there are essential informational elements (data bits) and nonessential control or padding information. We first examine the fields in a typical LAN frame. When expanding or establishing a network or when connecting disparate LANs, we will need to consider the overheads associated with the frame format.

ETHERNET

Preamble	Destination Address	Source Address	Type	Data	FCS
8 bytes	6 bytes	6 bytes	2 bytes	46 to 1500 bytes	4 bytes

IEEE 802.3

Preamble	Start of Frame Delimiter	Destination Address	Source Address	Length	Data	FCS
7 bytes	1 byte	6 bytes	6 bytes	2 bytes	46 to 1500 bytes	4 bytes

Figure 11.1   Ethernet and IEEE 802.3 Frame Formats

11.2.1   Ethernet Frames

Figure 11.1 depicts the standard composition of both IEEE 802.3 and Ethernet frames.

Note that the preamble in IEEE 802.3 is seven bytes, whereas in Ethernet it is eight bytes. Both are used for synchronization and consist of a repeating sequence of binary 1s and binary 0s. The IEEE 802.3 standard replaced the last byte of the Ethernet preamble field with a one-byte Start-of-Frame (SOF) delimiter. That byte has a sequence of 1s and 0s, but terminates with two set bits. Another difference between the two frames occurs in the protocol field in Ethernet that was replaced by the length field in the IEEE 802.3 frame. The two-byte protocol field contains a value that identifies the protocol transported in the frame, such as IP or IPX. In comparison, the length field identifies the length of the frame in an IEEE 802.3 environment. This means than only one protocol can theoretically be transported in an IEEE 802.3 frame. Because most organizations need to transport multiple protocols, the data field of the IEEE frame was used to convey several subfields that allowed multiple protocols to be transmitted. Referred to as a Subnetwork Access Protocol (SNAP), this frame retains the IEEE 802.3 frame format, but inserts special codes within the beginning of the data field to indicate the type of data transported.

Whereas some vendors produce dually functioning IEEE 802.3 and Ethernet hardware, this is done mostly to preclude the wholesale replacement of idiosyncratic (IEEE 802.3 versus Ethernet) NICs in their workstations. We now discuss the frame fields in order.

11.2.1.1 Preamble

The preamble field consists of alternating 1s and 0s that serve to announce the arrival of the frame and for all listeners in the network to synchronize themselves. Furthermore, this field serves to ensure a minimum 9.6-microsecond (µs) frame spacing at 10 Mbps to use for error control and recovery.

11.2.1.2 SOF Delimiter

This field, which only applies to IEEE 802.3 frames, consists of a format identical to the preamble, with alternating 1s and 0s for the first six bits. The seventh and eighth bits are both set to 1, which breaks the synchronization pattern and alerts the listener that the data is coming.

A controller strips off the preamble and SOF delimiters from incoming frames before buffering them. Accordingly, as the preamble and SOF delimiter are included in computations of length or they are not, minimum and maximum frame lengths can be determined differently in calculations. That is, minimum and maximum length frames will be eight bytes longer on the media than when in a computer's NIC.

11.2.1.3 Source and Destination Addresses

Both source and destination addresses occur in IEEE 802.3 and Ethernet frames. The destination address indicates the recipient of the frame and the source address indicates the originator. Two-byte source and destination addresses apply only to IEEE 802.3 and, although designed for use by small LANs, were never seriously implemented. In comparison, six-byte addresses apply to IEEE 802.3 and Ethernet and are *de facto* addressing standards. They exist within two special fields, as depicted in Figure 11.2. Those fields are:

- I/G (Individual/Group) bit — this is 0 for unicast frames and 1 for multicast frames
- U/L (Universal/Local) bit — applies only to six-byte addresses
  - 0 => universally assigned by IEEE
  - 1 => locally administered by vendor

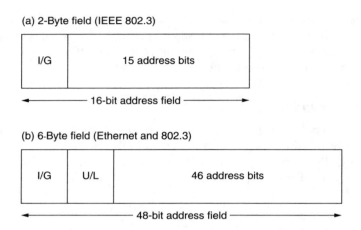

(a) 2-Byte field (IEEE 802.3)

| I/G | 15 address bits |

◄─────────── 16-bit address field ──────────►

(b) 6-Byte field (Ethernet and 802.3)

| I/G | U/L | 46 address bits |

◄─────────────── 48-bit address field ───────────────►

Figure 11.2   Source and Destination Address Field Formats

11.2.1.4   Type

The type field only applies to Ethernet. This field identifies the network layer protocol carried. There is a different connotation for IEEE 802.3 and therefore rules out interoperability between the two protocols.

11.2.1.5   Length

The length field is two bytes and is used to identify the number of data bytes in the data field. As noted, there is a proviso in length calculations according to whether the preamble and SOF are or are not included, but this does not affect the length field's value.

Short frames have an effect on reliable MAC delivery. It is possible that a short frame has collided and corrupted with another, but the sender still believes the transmission is successful. To preclude this possibility, it is deemed that the minimum length of all frames on an Ethernet must be at least twice the media's propagation delay. For instance, in a 10 Mbps coax-based LAN with a maximum length of 2500 m, the minimum time per the IEEE 802.3 standard is 51.2 μs. In turn, that time corresponds to 64 bytes, because 64 bytes $\times$ 8 bits/byte $\times$ 10^{-7} s/bit is 51.2 μs. As network speed rises, either the minimum frame length must also rise or the maximum segment length must fall.

11.2.1.6   Data Field

The data field has a minimum value of 46 bytes to ensure that the frame is minimally 72 bytes in length. The effect of having a minimum length data field requires that information that is less than 46 bytes be padded

SSD	Preamble	SFD	Destination Address	Source Address	L/T	Data	FCS	ESD
1 byte	7 bytes	1 byte	6 bytes	6 bytes	1 byte	46 to 1500 bytes	4 bytes	1 byte

Figure 11.3   Fast Ethernet Frame Format

to reach the minimum length. In certain documentation, this is referred to as a PAD subfield. Regardless of the manner, reference fill characters are added when necessary to ensure the minimum length of 46 bytes.

The maximum length of the data field is 1500 bytes. The implication of this is that data-intensive applications such as multimedia imaging and file transfers must use multiple frames.

11.2.1.7   Frame Check Sequence

The FCS field is four bytes in length. A Cyclic Reliability check (CRC) is calculated using both address fields, the type/length field, as well as the data field. This CRC is placed in the four-byte FCS field by the sender. The receiver then recalculates the CRC at the other end. If CRC sent and CRC received match, the frame is accepted; otherwise, the receiver simply drops the frame.

There are two other possibilities that can occur that will result in a frame being dropped. Those possibilities include:

1. Length of data field does not match the value in the length field.
2. Frame length is not a multiple of eight.

11.2.2   Fast Ethernet Frames

Figure 11.3 illustrates the Fast Ethernet frame in detail. We see that the frame format for Fast Ethernet is similar to the IEEE 802.3 frame, except for the Start of Stream delimiter (SSD) and End of Stream delimiter (ESD). SSD signals the arrival of a frame, whereas ESD indicates that the frame has been successfully transmitted.

The other thing about Fast Ethernet frames is that Ethernet and IEEE 802.3 are Manchester encoded with an interframe gap of 9.6 µs between frames.

In comparison, Fast Ethernet is transmitted using 4B5B encoding and an interframe gap of 0.96 µs. Both SSD and ESD fall within this gap.

11.2.3   Gigabit Ethernet Frames

Recall that the operating speed of an Ethernet network is reflected in either an increase in frame length or a decrease in maximum segment length.

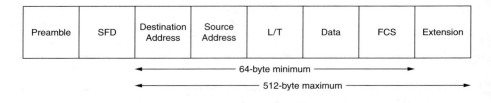

Preamble	SFD	Destination Address	Source Address	L/T	Data	FCS	Extension

◄———————— 64-byte minimum ————————►
◄———————— 512-byte maximum ————————►

Figure 11.4   Gigabit Ethernet Frame Format with Carrier Extension

At 1 Gbps, if a minimum frame length of 64 bytes is maintained, the network separation falls to 20 m. In structured cabling within an office building, horizontal cabling takes up to 10 m from the wall socket to the desktop. Therefore, to accomplish an increase of network cabling to around 200 m, two special techniques are employed:

1. Carrier extension
2. Packet bursting

These were addressed in detail in Chapter 10, but are quickly revisited here for the sake of continuity.

11.2.3.1   Carrier Extension

Carrier extension extends the Ethernet slot time to 512 bytes (from 64 bytes). This is achieved by padding the minimal 64-byte frame from the earlier 64 bytes. This action results in the carrier signal being placed on the network with an extension of up to 512 bytes. Figure 11.4 depicts Gigabit Ethernet with carrier extension. This results in an extension of the Ethernet slot time from 64 bytes to 512 bytes. To accomplish this, frames less than 512 bytes in length are padded with special carrier extension symbols.

At the receiver end, extension symbols are stripped off prior to FCS checks. Note that carrier extension only applies to half-duplex Ethernet. This significantly degrades performance, especially when coupled with short packets. Hence, packet bursting was introduced.

11.2.3.2   Packet Bursting

If a station has multiple frames to transmit, it does so after the first (padded) frame is successfully transmitted. Subsequent frames are not padded, but are limited by the maximum frame length. Figure 11.5 depicts Gigabit Ethernet with packet bursting.

We note therein that the first two packets transmitted were less than 512 bytes in length and were extended. Future packets within the burst

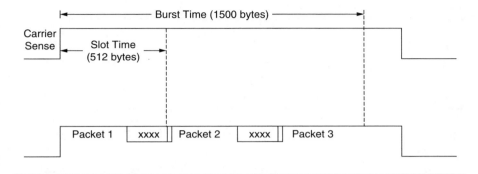

Figure 11.5   Gigabit Ethernet Packet Bursting

time of 1500 bytes are transmitted to completion. This is indicated by Packet 3. The interframe gap between frames is reduced from 9.6 µs on a 10 Mbps LAN to 0.096 µs on Gigabit Ethernet.

11.2.4   Frame Overhead

Table 11.1 summarizes the frame overhead percentage associated with transporting in Ethernet and Gigabit Ethernet frames as the number of bytes of information varies from 1 byte to 1500 bytes. As indicated, the percent overhead varies from 1.7 percent to 98.61 percent. We see that performance can degrade considerably with interactive traffic. The information in Table 11.1 can be important for network performance, especially in client/server situations, where it is preferable to send a lesser number of frames with information for several transactions at once, as that reduces the number of interframe gaps, which in turn improves the efficiency of the data flow.

11.3   AVAILABILITY LEVELS

We first define availability (A):

$$A\% = [\text{operational time/total time}] \times 100$$

expressed as a percentage. Consider a bridge that works round the clock. Over a year's time, assume that the bridge failed once and took 8 hours to repair. So out of 8760 hours per annum, the device was operational for 8752 hours. Then the availability of the bridge becomes:

$$A\% = (8752/8760) \times 100 = 99.9\%$$

Table 11.1   Frame Overhead

Bytes	Ethernet		Gigabit Ethernet	
Info in Data Field	Ratio of Overhead/ Frame Length	Percentage Overhead	Ratio of Overhead/ Frame Length	Percentage Overhead
1	71/72	98.61	519/520	98.61
10	62/72	86.11	510/520	98.08
20	52/72	72.22	500/520	96.15
30	42/72	58.33	490/520	94.23
45	27/72	37.50	475/520	91.35
46	26/72	36.11	474/520	91.15
64	26/90	28.89	456/520	87.69
128	26/154	16.88	392/520	75.38
256	26/282	9.22	264/520	50.77
512	26/538	4.83	26/538	4.83
1024	26/1050	2.48	26/1050	2.48
1500	26/1526	1.70	26/1526	1.70

There are two options to increase the availability of devices. Either deploy redundant devices or devices with multiple ports.

What are the implications?

Reliability is typically measured in terms of mean time between failures (MTBF) and mean time to repair (MTTR). Can these parameters be used to better understand availability levels? The answer can be obtained in the formula for availability expressed in terms of MTTB and MTTR as follows:

$$A\% = [MTBF/(MTBF + MTTR)] \times 100\%$$

From the above formula, it is important to remember that these are mean or average times. Otherwise the calculations will be erroneous. The mean times need to be measured across a range of devices installed; this is the MTBF information provided by device vendors, which may be used in place of the in-house determined average figures.

We note that if devices are connected in series, then for the system availability A_s:

$$A_s = \Pi \, A_i$$

Whereas, if devices are connected in parallel:

$$A_s = \{1 - \Pi \, (1 - A_i)\} \times 100$$

For hybrid topologies of devices, one can consider the system as a sequence of serial and parallel elements and compute the overall level of availability as simply as one computes the impedance of a block of series and parallel resistors.

11.4   NETWORK TRAFFIC ESTIMATION

We now consider the use of an a priori scheme to estimate or predict network performance. Plan for segmentation if high use is predicted, using a local bridge, switching device, or similar device to enhance performance.

In this case, one estimates traffic by considering the required functions of each network user. Group or classify similar network users into a workstation class. Do the calculations for one typical member of the workstation class, then multiply by the number of workstations in that class to obtain an estimate of traffic for the entire class. Repeat the procedure for all workstation classes and add the results to arrive at the average traffic for the entire network.

For a typical workstation class, activities performed may include:

- Load application
- Load graphic image
- Save graphic image
- Send e-mail message
- Receive e-mail message
- Print graphic image
- Print text data
- Invoke a client/server database

After selecting the activities, determine:

- Message size
- Number of frames per message
- Frame size
- Frequency per hour

Subsequently, use the following formula:

$$\text{Bit Rate} = (\text{frames/message} \times \text{frame size} \times 8 \times \text{frequency/hour})/(3600 \text{ s/hour})$$

Now that you have calculated this number for each activity performed for the workstation class, add up the bit rates (bps) for the entire class representative.

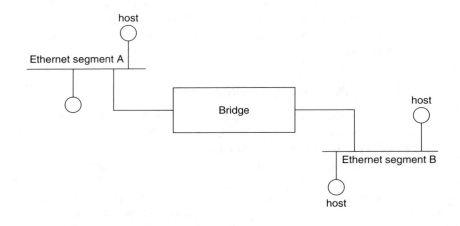

Figure 11.6 A Subdivided Network

If there are N stations in the class, then the bit rate per class = $N \times$ (bit rate for representative). Typical classes may include (project) managers, architects, secretaries, engineers, programmers, and system administration staff.

Finally, add the computed bit rates for all classes to arrive at the total estimated bit rate for the network. The projected growth rates for workstation classes may then be estimated to arrive at the projected bit rate for future utilizations.

When projecting the traffic load, it is important to note that utilization levels of Ethernet beyond 50 percent will result in performance degradation that begins to become observable. At such a time, you should consider segmentation using two-port local bridges, which are less expensive than routers or switches. Figure 11.6 illustrates this situation, by example, placing selected user classes within separate bridged segments.

For the example shown in Figure 11.6, let us assume that the most busy workstation class is programmers. Then, we may wish to consider placing all the users of that class in a separate segment connected to the other segments via a local bridge. If, for instance, the network utilization for the network is 65 percent, with the programmers consuming 54 percent, then segmenting the network results in a utilization of (65 × 54) percent or 35 percent for that class. Albeit not a perfect improvement, it is a betterment of the situation.

From this example, we can note why server farms should not all be placed in one segment. It is akin to putting all one's eggs in the same basket, with most if not all client/server transactions in the same segment.

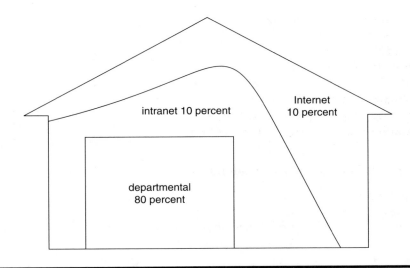

intranet 10 percent

Internet
10 percent

departmental
80 percent

Figure 11.7   Typical LAN Information Distribution

Figure 11.7 illustrates the 80-20 rule. When interconnecting two or more separate networks, this so-called 80-20 rule applies, with 80 percent of traffic typically intra-LAN and the remaining 20 percent inter-LAN.

11.5   AN EXCURSION INTO QUEUING THEORY

Now we need a procedure to estimate waiting times in the system and to select appropriate equipment with sufficient memory to meet specific requirements of the organization. To do so, we need some information about arrrivals and servicing times for frames arriving in a network. Then we apply classical results from queuing theory to arrive at characteristics of the system. Queuing theory has to do with managing both delays and buffer memory in remote bridges and routers used to link networks. If the delay is too high or the memory too low, performance in these devices degrades, necessitating retransmissions.

Queuing theory affords us models to determine and manage delays, to investigate the effect of modifying operating rates of circuits, as well as to determine the minimum acceptable memory requirements for devices to maintain a satisfactory level of performance. We now turn our attention to these features, as determined by queuing theory.

Consider the scenario where two LANs are interconnected via remote bridges or routers. Assuming a single-channel, single-phase queuing model with Poisson arrivals and arbitrary servicing times, the following formulas apply and yield information on waiting line characteristics:

λ = mean arrival rate
μ = service rate
Utilization: P = λ/μ
Probability: P0 = 1 - λ/μ
Length of queue = Lq = [$\lambda2$ $\sigma2$ + (λ/μ) 2]/2 (1 - λ/μ)
Length of system: L = Lq + λ/μ
Waiting time in queue: Wq = lq/λ
Waiting time in system: W = Wq + 1/μ

11.5.1   Buffer Memory Considerations

We assume that:

$$P_n = \text{probability of } n \text{ units in a system}$$
$$= (\lambda/\mu)^n (1 - \lambda/\mu)$$
$$= p^n (1 - p)$$

With:

$$p\,(n > k) = (\lambda/\mu)^k = p^k$$

Where:

$$\mu = \text{service rate}$$
$$\lambda = \text{arrival rate}$$

For:

$$n = 0 \text{ to } n = 20$$
$$k = 0 \text{ to } k = 20$$

Table 11.2 depicts these probabilities.

Using the data in Table 11.2, you can make the following computations.

To obtain a level of 99.9 percent of the occurrences, which is equivalent to saying 0.1 percent of nonmanageable occurrences, note first from the table that:

When k = 7,
P (N > 7) is 0.00104284

So one selects k = 8 to satisfy the requirement for handling 99.9 percent of occurrences, wherein the frame arrival rate exceeds the servicing rate of the bridge or router.

Table 11.2   Probabilities

Probability of N Units		Probability of K or More Units	
N	P(N)	K	P (N > K)
0	0.62500000	0	1.00000000
1	0.23437500	1	0.37500000
2	0.08789063	2	0.14062500
3	0.03295898	3	0.05273438
4	0.01235962	4	0.01977539
5	0.00463486	5	0.00741577
6	0.00173807	6	0.00278091
7	0.00065178	7	0.00104284
8	0.00024442	8	0.00039107
9	0.00009126	9	0.00014665
10	0.00003437	10	0.00005499
11	0.00001289	11	0.00002062
12	0.00000483	12	0.00000773
13	0.00000181	13	0.00000290
14	0.00000068	14	0.00000109
15	0.00000025	15	0.00000041
16	0.00000010	16	0.00000015
17	0.00000004	17	0.00000006
18	0.00000001	18	0.00000002
19	0.00000001	19	0.00000001
20	0.00000000	20	0.00000000

P (N units) and P (K+ units in system)

P (N frames in system)

So, if frame length is 1200 bytes, the memory requirement is:

$$\text{Memory} = 1200 \text{ bytes/frame} \times 8 \text{ frames}$$
$$= 9600 \text{ bytes}$$

This procedure yields a nine-step approach for determining storage requirements:

1. Set λ = mean arrival rate
2. set μ = mean servicing rate
3. Determine the utilization level
4. Determine the level of service required when $\lambda > \mu$

5. Set N = units in system
6. Set K = level of service the server is required to provide
7. Find p(N > K)
8. Extract K = number of frames to be queued
9. Multiply average frame length by the number of frames *K* to be queued

This yields the memory values for all situations with a predetermined probability level.

11.6 ETHERNET PERFORMANCE DETAILS

Because of the random nature of collisions in the CSMA/CD protocol, Ethernet bus performance is nondeterministic. As a result, performance characteristics and delays are not predictable. All we have are average and peak utilizations and this is information we may use to segment an existing network to enhance performance.

11.6.1 Network Frame Rate

This parameter for Fast Ethernet is ten times the value for 10 Mbps Ethernet. Similarly, the rate for Gigabit Ethernet is ten times that for Fast Ethernet, but that proposition is true only for certain types of frames, when carrier extension is used.

Let us quickly revisit the IEEE 802.3 frame formats:

- Preamble (8 bytes) [7 byte preamble and 1 byte SOF delimiter]
- Destination address (6 bytes)
- Source address (6 bytes)
- Length/type (2 bytes)
- Data (46 to 1500 bytes)
- FCS (4 bytes)
- Total = 72 bytes to 1526 bytes

Under Ethernet and IEEE 802.3 frames operating at 10 Mbps, there is a dead gap of 9.6 μs between frames. This can be used to determine the frame rate on the network. For example, consider a 10 Mbps LAN.

The bit time is 10^{-7} s or 1 ns.

If we assume frame length of 1526 bytes maximum, then the time per frame is:

$$9.6 \text{ μs} + 1526 \text{ bytes} \times 8 \text{ bits/byte} \times 100 \text{ ns/bit} = 1.23 \text{ ms}$$

Table 11.3   Ethernet Frame Processing (Frames per Second)

Network Type	Average Frame Size (Bytes)	Frames per second	
		50% Load	100% Load
Ethernet	1526	406	812
	72	7440	14,880
Fast Ethernet	1526	4060	8120
	72	74,400	148,800
Gigabit Ethernet	520	117,481	234,962
	1526	40,637	81,274

Because one 1526-byte frame requires 1.23 ms, in 1 s there are 1/1.23 or approximately 812 maximally sized frames. Such a situation using maximal frames occurs when doing data intensive file transfers.

Table 11.3 shows frame rates for Ethernet, Fast Ethernet, and Gigabit Ethernet. It summarizes the frame processing requirements for these networks under 50 percent and 100 percent load conditions, based on minimum and maximum frame sizes. These rates indicate the number of frames that a bridge connected to a LAN must be capable of handling.

If a bridge, switch, or router is used on a LAN, the data contained in Table 11.2 can be used to determine the minimum required processing speed for the device.

11.6.2   Gigabit Ethernet Considerations

In Gigabit Ethernet, carrier extension is used to ensure a minimum frame length of 512 bytes (or 520 with preamble or SOF delimiter). The carrier extension runs from 0 to 448 bytes according to the length of the pure data content of the frame. The interframe gap is 0.096 μs.

This in turn entails that:

Frame Rate = 0.96 μs + 520 bytes × 8 bits/byte × 1 ns/bit = 4.256 μs

So, in 1 s there are a maximum of:

1/4.256 = 234,962 minimum sized frames

In turn, this means that with the transmission of a maximum length 1526-byte frame, the frame rate is:

0.096 μs + (1526 byte × 8 bits/byte × 1 ns/bit) = 12.304 μs

Therefore, in 1 s there can be a maximum of 81,274 maximum length frames.

If we look again at Table 11.3, we see easily that the performance ratios of Fast Ethernet and Gigabit Ethernet are actually 1.579:1 and 15.79:1, respectively, in relation to 10 Mbps Ethernet. This is despite the 10:1 and 100:1 increase in data rates.

11.6.3   Actual Operating Rate

To estimate the actual operating rate of an Ethernet network, you need to deduct the dead time from the maximum throughput. For example, at 10 Mbps our computation would be as follows:

$$10 \text{ Mbps} - (9.6 \text{ } \mu s/100 \text{ ns} \times 812) = 9,922,048 \text{ bps} = 9.922 \text{ Mbps}$$

Therefore, we see that for maximum utilization of an Ethernet, one must transmit 9.922 Mbps of the actual data rate.

Similarly, for Fast Ethernet:

$$\text{Operational Rate} = 100 - (.96 \text{ } \mu s/10 \text{ ns} \times 8127)$$
$$= 99.22 \text{ Mbps}$$

For Gigabit Ethernet, the operational rate is 992.19 Mbps.

11.7   BRIDGING A NETWORK

When utilization levels reach 50 to 60 percent for extended time periods, LAN modification is in order. One way to do this is to use local bridges to segment the bigger LAN. Once the decision has been made to bridge the network, the next step is to determine that the filtering and forwarding rate of the bridge is minimally equal to the actual operating rate (frames/sec) of the network as measured or estimated.

Until now, we assumed that two bridged segments have the same operating rates. This is not always the case. For example, one department may be working at 10 Mbps, whereas another uses Fast Ethernet. What is the throughput between the LANs? Figure 11.8 below depicts this situation.

To compute the time to transfer information from one network to the other, let us assume the following parameters:

$$R_T = \text{time taken to send A} \rightarrow B$$
$$R_1 = \text{time taken to send A} \rightarrow Br$$
$$R_2 = \text{time taken to send Br} \rightarrow B$$

Figure 11.8   Linking LANs with Different Operating Rates

Then, the time to transmit data between the two networks becomes:

$$R_T = (R_1 \times R_2)/(R_1 + R_2)$$

For instance, if we estimated:

$$R_1 = 812 \text{ frames/s @ 10 Mbps}$$
$$R_2 = 8130 \text{ frames/s @ 100 Mbps}$$

Then:

$$R_T = 738 \text{ frames/s}$$

assuming that the sending station has full access to the bandwidth and the resources of the bridge.

12

ISSUES AT THE NETWORK, TRANSPORT, AND APPLICATION LAYERS

12.1 INTERNETWORKING OVERVIEW

We begin our discussion of issues at the upper layers, with an overview of internetworking at large. We introduced the router in Chapter 8 and noted that some of its key functions include the linking of different networks, routing and delivery of data between processes and applications in End Systems (ESs) an ISO terminology for edge devices on different networks, and to do all this seamlessly and transparently in relation to the network architecture in the attached networks.

One protocol, and predominantly the popular one, that supports these functions is the IP or IPv4. Figure 12.1 shows the IP header, which is a minimum of 20 octets. The fields in this depiction are as follows:

- Version (4 bits) — indicates version number to enable evolution.
- Internet header length (IHL) (4 bits) — length of the header in units of 32-bit words. Minimum value of IHL is 5 and maximum is 20 octets.
- Type of service (8 bits) — this yields guidance to ES IP modules and to routers en route about the relative priority of packets.
- Total length (16 bits) — total IP packet length in octets.
- Identification (ID) (16 bits) — a sequence number that, along with the source or destination addresses and user protocol, identifies a packet uniquely. Given these three values, this ID field is then unique in value.
- Flags (3 bits) — only two bits are defined here. The more bit indicates whether this is the last fragment in the original packet. The do not

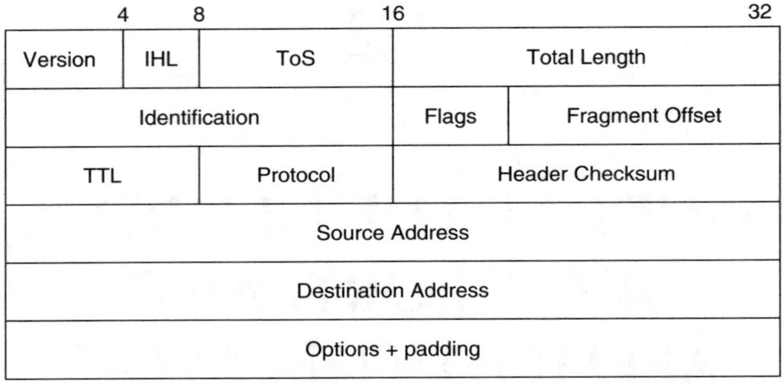

	4	8	16	32

Version	IHL	ToS	Total Length	
Identification			Flags	Fragment Offset
TTL		Protocol	Header Checksum	
Source Address				
Destination Address				
Options + padding				

Figure 12.1   The IP Header

fragment bit prohibits further fragmentation when set. Note that, if en route, the second flag bit is set and if the packet exceeds the maximum transmission unit (MTU) size, the packet is simply discarded. Hence, source routing is preferable when this bit is set to avoid subnetworks with too low of a MTU size.

■ Fragment offset (13 bits) — this field indicates where in the original packet this fragment belongs, in 64-bit units. By implication, all but the last packet must contain a data field in multiples of 64 bits in length.

■ Time to live (TTL) (8 bits) — how long, in seconds, a packet is allowed to remain on the internet. Each router en route to the destination must decrement the TTL by one at least, so this field is similar to a hop count.

■ Protocol (8 bits) — indicates the next higher layer protocol, which is to receive the data in the destination, essentially indicating the type of the earlier header in the packet after the IP header.

■ Header checksum (16 bits) — an error detection code applied only to the header. Because some header fields change in transit, this checksum is recomputed and reverified at each router en route.

■ Source address (32 bits) — formulated to allow a variable allocation of bits to specify the network and ES attached to the specified network.

■ Destination address (32 bits) — same as source address.

■ Options (variable) — encodes requested options such as security, source routing, routing recording, and timestamps.

■ Padding (variable) — used to ensure that packet length is a multiple of 32 bits in length.

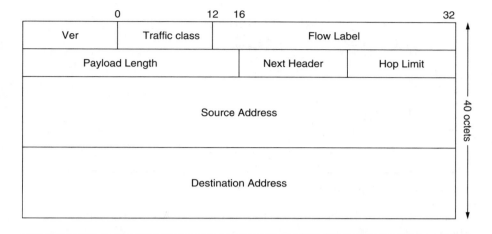

Figure 12.2   The IPv6 Header

We observe that a new standard for IP addressing was initially specified by the Internet Engineering Task Force (IETF), variously called IPv6 or IPng. This scheme is depicted in Figure 12.2, where we see that the IPv6 uses addresses 128 bits in length.

IPv6 supports the higher speeds of today's networks and the mix of multimedia data streams. Basically, there was a need for more addresses to assign to all conceivable devices. As noted, the source and destination addresses are 128 bits in length. It is expected that all TCP/IP installations will eventually graduate to IPv6, although this process may take many years, if not decades, to be achieved.

12.2   PROTOCOL ARCHITECTURE

Figure 12.3 shows two LANs interconnected over an X.25 network. Therein, we see the operation of IP for data exchange between ESs, A and B, attached to the two LANs.

The IP at A receives blocks of data from the higher layer protocol software at A. It then attaches an IP header with the global IP address of B. This address consists of a network ID and an ES identifier and the resulting unit is called an Internet Protocol data unit (IPDU) or simply a datagram.

The datagram is encapsulated within the LAN protocol and sent to Router X, which promptly decapsulates the LAN fields to examine the IP header. The same router then further encapsulates the datagram with X.25 protocol fields and sends it across the WAN to the remote Router Y. That router then decapsulates the X.25 fields and recovers the datagram, then

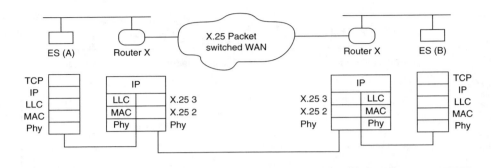

Figure 12.3 IP Operation

wraps it up with Layer 2 fields as suitable for LAN 2 and sends it off to the Device B.

12.3 DESIGN ISSUES

We now turn our attention to some design issues in more detail. These include:

- Addressing
- Routing
- Datagram lifetime
- Fragmentation or reassembly

12.3.1 Addressing

A unique address is associated with each ES and intermediate system (IS) router within a configuration. This is called an IP address and is used to route a datagram through an internet to the target system indicated by the destination address.

Once data arrives at the remote ES, it must be processed and delivered to some process or application therein. Typically, multiple applications will be concurrently supported and one application may support several users.

Each application and maybe each user of an application is assigned in the system architecture with a port. Minimally, each application has a port number that is unique in that system. Furthermore, for instance, a File Transfer Protocol (FTP) application may support several concurrent data transfers and in that situation each transfer is dynamically assigned a unique port number.

There are two levels of IP addressing. First, there is the globally applicable (and possibly redundant) IPv4 address. Second, for each device interface on the network, there is a unique address. Examples are the MAC address (802 network) and the X.25 host address. These are sometimes referred to as network attachment point addresses (NAPAs).

The issue of addressing scope is relevant only for global IP addresses. On the other hand, port numbers are unique only within a given system. Hence if there are two systems, A and B, the following ports work uniquely, i.e., A.1 and B.1.

12.3.2 Routing

Generally, routing is achieved by maintaining a table within each router. The routing table yields, for a given target system, the next hop (next router) to which the datagram should be sent.

Routing tables can be static or dynamic. Whereas a static table can contain redundant routes for routers that are not available, a dynamic routing table is more flexible. "Neighbor greeting" is used to determine the next hop. This helps with congestion control and also to address the mismatch between LAN and WAN transmission rates.

Source routing occurs when a sending station dictates a sequential list of routers that the datagram must traverse; this specification is inserted within the datagram itself by the sending station. Route recording is when each router in the trajectory of the datagram appends its IP address to the datagram. This often helps with network maintenance and troubleshooting.

12.3.3 Datagram Lifetime

There is a possibility, especially with dynamic routing, that a datagram or some of its fragments keeps circulating endlessly in the internet, especially when there are sudden, significant changes in the network traffic or when there is a flaw in the system's routing tables. To preclude this problem, datagrams are sometimes marked with a lifetime field similar to a hop count. The hop count is set initially to N and decremented by one as the datagram passes through each router en route. When the hop count reaches zero, the datagram is discarded.

12.3.4 Fragmentation or Reassembly

Subnetworks within an internet may specify different MTUs, which refers to the largest size of datagrams in transit. It is not feasible to dictate one uniform maximum packet size across networks. So when the next subnetwork has a smaller MTU size compared to the previous subnetwork, the only option is to fragment the packet. (Of course, unless the do not fragment bit flag is set, then the packet is discarded).

In IP, reassembly of fragmented datagrams happens in the ES at the destination. The following fields in the IP header are used for handling fragmentation and reassembly:

- Data unit identifier (ID) — a composite of source and destination addresses, an identifier of the higher layer protocol that generated the data (e.g., TCP), and a sequence number supported by that protocol layer
- Data length — the length of the user data field
- Offset — the position of a fragment of user data in the data field of the original datagram in multiples of 64 bits
- More flag — specifies that more fragments follow

One problematic issue that must be dealt with is that of lost fragments. IP does not guarantee delivery. There are two mechanisms to deal with this issue.

First, one mechanism is for the reassembly function to generate a local real-time clock that keeps ticking. If the clock expires prior to reassembly, that entire effort is abandoned and received fragments are discarded.

A second approach uses the datagram lifetime, a part of the header of each incoming fragment. The lifetime field continues to be decremented by the reassembly function. If the lifetime expires prior to full reassembly, received fragments are discarded.

12.4 ROUTING AND ROUTE PROTOCOLS

In essence, a router is primarily a packet switch. In this connection, the term packet refers to the so-called protocol data unit (PDU) that traverses from the Network Layer 3 software in one system across a network to the Layer 3 software in another system. The packet contains, among other information, the addresses of source and target ESs. The ESs are the devices that generate or receive the overwhelming majority of all packets traversing the network.

Another piece of terminology is the "subnetwork." This refers to a collection of network resources that can be reached without going through a router. If a router is involved in going through the network from one ES to another, these ESs are in different subnetworks.

Finally, an internetwork (or simply an internet) refers to a collection of two or more subnetworks interconnected by routers.

In this context, the role of the router is simple. It receives packets from ESs (and possibly other routers) and routes them through the internet to the appropriate destination network. Once a packet has arrived at the destination subnetwork, the last router in the path traversed forwards the packet to the intended ES recipient.

Some of the other more advanced functions possibly supported by routers include fragmentation, congestion control, and fairly sophisticated packet filters providing a modicum of security.

The frame is the mechanism to transport data. Packets ride within frames and are the PDU to get from one ES to another across an internet. If two ESs are within the same subnetwork, the packet is framed by the transmitter and can pass through bridges. If the two ESs are in different subnetworks, the packets will go from the sender to the local router, which decapsulates the Layer 3 header and sends it across to the next IS in a new frame.

It is worth mentioning that whereas bridges maintain forwarding tables and track virtually all devices on the subnetwork, there are no entries in these tables for ESs on different subnetworks and how to reach them. In the latter case, data must pass through routers.

12.5 ROUTING REVISITED

The network layer protocol in an internet is responsible for all end-to-end routing of packets. There are many such protocols and they all share some common features. They all use a packet structure and an address format. All of them specify a network ToS, as well as other issues like fragmentation, connectionless versus connection-oriented service, and packet prioritization.

The fundamental concept for the network layer protocol is the Layer 3 address. This address is hierarchical, with at least two addressing segments defined. The first identifies a subnetwork and the second identifies an ES within that subnetwork. These two fields are always present, whereas some specific Layer 3 protocols define additional fields.

Note that each router and device interface must have a unique Layer 3 address, a situation not unlike an area code and local phone number in the telephone industry. This unique IP address is the basis for all routing within the internet.

The vast majority of all Layer 3 protocols are connectionless and datagram (best-effort) based. We note that a connectionless service is one wherein the upper layer protocol or architecture has no means of requesting an end-to-end relationship or connection with another ES. All that the Layer 3 protocol can do is to provide data with a destination address. All acknowledgments, flow control, and sequencing of messages are managed by the upper layer protocol or applications.

A datagram network is one wherein routers are unable to establish an end-to-end circuit for traffic. Every packet received by a router is routed independently of earlier or later packets. So no guaranteed QoS can be provided. On the other hand, if the network supports end-to-end circuits,

Table 12.1 Major Network Layer Route Protocols

Network Layer Protocol	Address Length (octets)	Address Fields (octets)	Additional Capabilities	Used In
Internet Protocol (IP)	4	NETID (var) HOSTID (var)	Fragmentation, nondelivery notice subnetting	Internet, most network environments
Internetwork Packet Exchange Protocol (IPX)	12	Network (4) Node (6) Socket (2)	Automatic client addressing	NetWare
Datagram Delivery Protocol (DDP)	4	Network (2) Node (1) Socket (1)	Automatic client addressing	AppleTalk
VINES Internet Protocol (VIP)	6	Network (4) Subnetwork (2)	Fragmentation, nondelivery notice, Automatic addressing	VINES

intermediate routers would know that packets will arrive on an established circuit and expected loads can be defined at the time of circuit setup.

Table 12.1 summarizes four of the major network layer protocols in use today along with their key features. All four are connectionless and datagram based.

IP addressing is the most complex of all. The boundary between the IP subnetwork number (NETID) and the ES number (HOSTID) is not rigidly fixed. The boundary varies depending on the address class and the subnet mask being used. In Table 12.2, we see that there are five address classes, three of which are used for deploying subnetworks.

Table 12.2 IP Addressing Overview

Address Class	1st Octet Value	Length of NETID	# of NETID	# of HOSTID
Class A	1–126	1 octet	126	16,777,214
Class B	128–191	2 octets	16,382	65,534
Class C	192–223	3 octets	12,097,150	254
Class D	224–239	Multicast	N/A	N/A
Class E	240–255	Reserved	N/A	N/A

The more difficult aspect of the IP addressing mechanism is the notion of the subnet mask. The purpose of a subnet mask is to take a NETID and divide it into smaller subnetworks connected by routers. For instance, the Class B address 128.13.0.0 can be subdivided into 256 smaller networks designated as 128.13.1.0, 128.13.2.0, 128.13.3.0, and so on, using the 255.255.255.0 subnet mask. This process is called subnetting. The mask may also be used by routers to summarize routes. For example, all the Class C networks from 199.12.0.0 to 199.12.255.0 can be advertised as 199.12.0.0 using the 255.255.0.0 mask. This process is called supernetting or Classless Inter Domain Routing (CIDR).

Usually IP addresses are assigned to devices manually, but there can be exceptions. For example, the Dynamic Host Configuration Protocol (DHCP) permits a DHCP server to dynamically lease addresses to ESs when they come online. IP can fragment packets if required, with reassembly at the destination ES.

12.5.1 Routing Protocols

Routing protocols are responsible for maintaining routing tables dynamically. The routing protocol monitors the network and accordingly updates the routing tables when network changes occur. Most Network Layer Route protocols use at least two routing protocols.

These protocols can be evaluated according to a number of criteria:

- Bandwidth
- Metrics
- Convergence time
- Memory space
- Processing power

Bandwidth is the first criterion for appraisal. Maintaining routing tables means that routers need to greet each other, which consumes bandwidth. The more bandwidth so consumed for administrative reasons, the less that is available to the protocol for its really intended purpose.

The next point is the metric that the routing protocol minimizes. Some use simple hop count, whereas other more sophisticated protocols use such metrics as delay, bandwidth, packet loss, or a combination of these metrics.

The third assessment point is convergence time. This is the delay between the occurrence of network changes and the time taken for all routers to refresh and update themselves with the most current state of affairs, as well as to alter the routing tables.

Finally, routing protocols use up memory space and processing power within the routers. With ever more powerful devices, this becomes less of an issue.

Regardless of the specific routing protocol used, these protocols can all be clubbed under the banner of distributed protocols meaning that route recalculation occurs at all routers in the internet. We note that centralized routing protocols wherein a single system makes all routing decisions and then downloading routing tables for all routers is a relatively recent phenomenon and is happening by degrees. There are two distinct types of distributed routing protocols:

1. Distance Vector (DV)
2. Link State (LS)

12.5.2   DV Protocols

DV protocols are also called Vector Distance or Bellman-Ford protocols. They have three important features:

1. Routing updates produced have a list of target or cost pairs
2. Updates are sent to all neighboring devices
3. Rerouting calculations are performed within each system

In essence, DV protocols extract a list of learned destinations and the costs to them and pass this knowledge on to neighboring devices. These neighbors then use this information to identify better routes than currently exist in routing tables, at which point the latter are updated.

Earlier in this book, we noted DV protocols have a count-to-infinity problem. Figure 12.4 illustrates this situation. We see therein that Router A has a route to Subnetwork 1 and Router B uses Router A to reach this subnetwork.

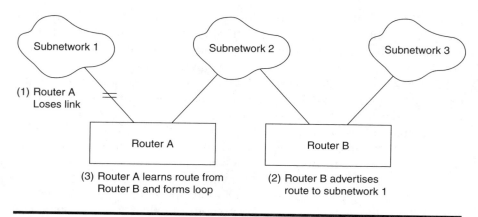

Figure 12.4   The Count-to-Infinity Problem

If Router A loses the route and Router B advertises this fact before Router A can advertise the loss, Router A will accept Router B as a new route to Subnetwork 1, forming a loop. Any packet destined for Subnetwork 1 and arriving at one of these routers will endlessly thrash between them.

To address this problem, most DV protocols employ a split horizon. This is a rule that prevents a router from advertising routes downstream, from where they were learned.

Split horizon with poison reverse is a version that permits advertisement of these routes but sets the cost to infinity, to prevent other routers from learning. Most DV protocols send complete updates periodically. This can seriously affect performance in relation to convergence times, so an improvement to this is to have event-driven (triggered) updates. However, if the table update was due to a failed (suboptimal) route, it is possible for a router that has not received information about the change to reintroduce the old route back into the network. To preclude this occurrence, routers are required to place failed routes into a hold down state, for a time typically three times the normal update interval. The first news (good or bad) travels quickly, but subsequent good news (new routes) is learned more slowly.

In summary, DV protocols are simple to design and implement with little demand for memory and processing power. However, convergence is a major problem and can consume a fair amount of network resources. Table 12.3 summarizes the key attributes of the major DV protocols, including RIP, RIP II, IGRP, EIGRP, RTMP, RTP, and BGP 4.

12.5.3   LS Protocols

The other family of routing protocols, LS, has three distinguishing features:

1. Routing update broadcast
2. LS database
3. Full route recalculation

Routing updates are broadcast, in a manner similar to flooding. Full route recalculation is performed last for all routers running the protocol. The central feature of LS protocols is the LS database. Routing updates as flooded are all stored in a local database. This database has enough information to graph out the entire network, calculate alternative paths, and construct a routing table. All the databases must synchronize. New routers in the network can obtain a copy of the database from nearby routers, whereas existing routers periodically revisit and verify the integrity of the database. The size of the database is impacted by the size and complexity of the network. Although updates are kept small and are event

Table 12.3 The Major Distance Vector Protocols

Routing Protocol	Used to Route	Metric(s)	Update Interval	Documented by
Routing Information Protocol (RIP)	Internet Protocol (IP)	Hop count	30 s	RFC 1058
RIP v2 (RIP II)	IP	Hop count	30 s	RFC 1388
Routing Information Protocol Exchange	Internet Packet Exchange (IPX)	Delay, hop count	60 s	Novell/Xerox
Interior Gateway Routing Protocol (IGRP)	IP	Delay, bandwidth (reliability, load)	90 s	Cisco
Enhanced IGRP (EIGRP)	IP, IPX, DDP	Delay, bandwidth (reliability, load)	Event driven	Cisco
Routing Table Maintenance Protocol (RTMP)	Datagram Delivery Protocol (DDP)	Hop count	10 s	Apple Computer
Routing Table Protocol (RTP)	VINES Internet Protocol (VIP)	Delay	90 s	Banyan
Border Gateway Protocol version 4 (BGP 4)	IP	Hybrid (also policy based)	Event driven	RFC 1771

driven, the flooding still consumes bandwidth. So there came about certain techniques to reduce the effect of this flooding of information.

First, only selected routers are required to send a flooded update, priorly checking the update against their database to avoid resending already sent updates. Another mechanism is for networks running LS to be segmented into areas, with updates flooded only within the areas. To handle interarea routing information, specific routers are assigned to summarize interarea routes to other areas.

LS protocols converge more quickly. They can be more bandwidth friendly than DV and less susceptible to routing loops. However, they can be more complex to design, configure, and implement. They also consume more router resources like memory and processing power.

Table 12.4 The Major Link State Protocols

LS Routing Protocol	Used to Route	Metric(s)	Documented by
Open Shortest Path First (OSPF)	IP	Dimensionless	RFC 2178
Intermediate System to Intermediate System (IS-IS)	IP, CLNS	Dimensionless	RFC 1142 & ISO DP 10589
Netware Link Services Protocol (NLSP)	IPX	Dimensionless	Novell

The Internet is the collection of all existing internets. Each internet is locally administered and referred to as an autonomous system (AS). Any routing protocol working within an AS is called an Interior Gateway Protocol (IGP). However, there is also a need to route across ASs when such protocols are called EGPs (Exterior Gateway Protocols).

Table 12.4 summarizes the key attributes of the major LS routing protocols that exist today, including OSPF, IS-IS, and NLSP.

12.6 EXCURSION INTO THE TRANSPORT LAYER

TCP/IP has two fundamental transport layer protocols:

1. TCP
2. User Datagram Protocol (UDP)

TCP is a reliable, two-way byte stream protocol and is rather complex. It guarantees in-sequence and accurate delivery of data by building a checksummed virtual circuit connection over and on top of IP's unreliable, connectionless best-effort service. It also deploys flow control and congestion control mechanisms that allow for efficient use of bandwidth. In this context, a window of packets is sent pending acknowledgment. This windowing mechanism is the basis for TCP's flow and congestion management capabilities.

TCP also supports multiplexing, whereby messages can be sent to different processes on the same host. It does this by means of port abstraction, wherein every process on an ES is assigned a unique (locally) port number. The TCP header has a source and destination port number. Services such as Telnet use this port abstraction and allow multiple clients to connect to the same service.

UDP is a best-effort unreliable connectionless protocol for applications without any sequencing/flow control requirements. It is often used when promptness, rather than accurate delivery, is sought — for example, when sending speech or video.

12.7 MULTIMEDIA SERVICE

We now turn our attention to the transport of multimedia applications, which is an instance of issues at the application layer.

VoIP is the enabling technology for multimedia service integration. Savings from integration can be significant and the technology is available.

Companies with a private voice network retain one or more PBXs to implement the integrated service. A PBX is the switching element linking two users in a voice or video connection. PBXs have three basic components:

1. Wiring
2. Hardware
3. Software

Wiring is dedicated to each phone in use. This allows employees to call each other. To gain access to the PSTN, PBXs need phone lines purchased from the telephone company.

Hardware includes a switched network connecting two phones and servers for PBX software. Software controls such functions as call setup, forwarding, call transfer, call hold, as well as generating per-call statistics. Organizations with multiple locations require PBXs dedicated at each site connected by public leased lines.

Moving toward full VoIP deployment, we will gradually see PSTNs replaced by public data networks, the introduction of VoIP gateways as well as VoIP gatekeepers, and the customer premise equipment (CPE). To fully support call management, the International Telecommunications Union's ITU-T standardized the H.323 protocol, which describes terminals, equipment, and services for multimedia connections over a LAN. Voice is only one of the services supported.

The H.323 recommendations are proving to be the basis by which many backbone, access, and CPE vendors are developing VoIP components with ensured interoperability. H.323 VoIP products can be broken down into the the following categories, mapping loosely to network layers:

- CPE — includes such devices as Microsoft NetMeeting® conferencing software, Intel® ProShare® conferencing software, the Selsius® Ethernet phone, as well as Symbol® NetVision® phone, an H.323 telephone that plugs into an Ethernet port.
- Network infrastructure equipment — includes standard routers, hubs, and switches. Because voice is sensitive to delays and losses, a number of router features like Random Early Detection (RED), weighted fair queuing (WFQ), Resource Reservation Protocol (RSVP),

compressed RTP, and multiclass, multilink PPP have evolved over the years to address these issues.

- Servers — provide a major VoIP benefit of utilizing the Internet model, with clear demarcation between network infrastructure and network applications. Servers support the applications. The H.323 gatekeeper service, for example, supports call control; an authentication, authorization, and accounting (AAA) server provides billing and accounting; and a Simple Network Management Protocol (SNMP) server provides for network management.
- Gateways — represent an evolutionary step for organizations moving toward VoIP. Given that it will be a while for the data network to handle all multimedia communication, gateways are an interim measure to link the new VoIP services with existing public or private voice networks.

12.8 SOME DELAY TIME CALCULATIONS

In this section, we turn our attention to computing the delay times and latency that occur when we deal with the transmission of small pieces of multimedia across an internet. Because UDP is used to trasnsport digitized voice, we first need to note that the UDP header is 16 bytes in length.

12.8.1 10 Mbps Ethernet, 100 Mbps Fast Ethernet, and 1000 Mbps Gigabit Ethernet

The delay time calculations for the 10 Mbps Ethernet, 100 Mbps Fast Ethernet, and 1000 Mbps Gigabit Ethernet are presented in Table 12.5.

Table 12.5   Some Ethernet Delay Calculations

Ethernet	Interframe Gap (μs)	Delay Time	Maximum Delay Time
10 Mbps Ethernet	9.6	$\Delta = 9.6\ \mu s + (8 + 6 + 6 + 2 + 20 + 16 + 100 + 7)$ bytes $\times\ 8$ bits/byte $\times\ 10^{-7}$ s/bit	$\Delta_{max} = 9.6\ \mu s + 1500$ bytes $\times\ 8$ bits/byte $\times\ 10^{-7}$ s/bit
100 Mbps Fast Ethernet	0.96	$\Delta = .96\ \mu s + (8 + 6 + 6 + 2 + 20 + 16 + 100 + 7)$ bytes $\times\ 8$ bits/byte $\times\ 10^{-8}$ s/bit	$\Delta_{max} = .96\ \mu s + 1500$ bytes $\times\ 8$ bits/byte $\times\ 10^{-8}$ s/bit
1000 Mbps Gigabit Ethernet	0.096	$\Delta = .096\ \mu s + (8 + 6 + 6 + 2 + 20 + 16 + 100 + 7)$ bytes $\times\ 8$ bits/byte $\times\ 10^{-9}$ s/bit	$\Delta_{max} = .096\ \mu s + 1500$ bytes $\times\ 8$ bits/byte $\times\ 10^{-9}$ s/bit

Key parameters for the calculations include the interframe gap (varies, given in µs), the MAC level preamble (8 bytes), destination MAC address (6 bytes), source address (6 bytes), type/length field (2 bytes), and the data field (minimum 46 bytes and maximum 1500 bytes). Within the data field is the encapsulated IP header. The entire 20 bytes of the IP header need to be read to get to the UDP header, because the UDP port must also be read. We also factor in (roughly) 7 bytes of RTP information, plus a small quantity (approximately 100 bytes) of multimedia data. For a maximum length Ethernet field, the frame length increases to 1500 bytes.

Note that the maximum delay time only applies to data transported between multimedia carrying packets. This explains why file transfers and similar operations that have frames inserted between two voice packets can cause distortion on slow Ethernet LANs.

By computing multimedia delay times, as well as the effect of the insertion of packets transporting data between digitized voice packets, it becomes possible to determine if your LANs can handle multimedia prior to implementing a new application.

12.8.2 Switches

For a store and forward Ethernet switch, the entire frame must be received, stored, and processed.

In this context, we can compute minimum and maximum delay times for a 10 Mbps switch as follows:

$$\Delta_{min} = 9.6 \ \mu s + (72 \ \text{bytes} \times 8 \ \text{bits/byte} \times 10^{-7} \ \text{s/bit})$$
$$= 67.2 \ \mu s$$

$$\Delta_{max} = 9.6 \ \mu s + (1526 \ \text{bytes} \times 8 \ \text{bits/byte} \times 10^{-9} \ \text{s/bit})$$
$$= .012304 \ s$$

Similar considerations apply for switches at higher speeds and computations may be made with the appropriate interframe gap and processing speeds at the higher rates.

When examining vendor specifications for latency, one needs to be cautious. Sometimes vendors do not denote the frame length used for latency measurements. This frame length has been factored into our computations carried out above.

13

WIRELESS LOCAL AREA NETWORKS

13.1   INTRODUCTION

Wireless LANs (WLANs) represent a major leap in the field of networking. From shipments of a few thousand network adapters during the turn of the millennium, sales have literally exploded. Millions of network adapters and access points are expected to be marketed during 2004.

There are numerous benefits associated with the use of WLANs. Notebook and laptop PCs equipped with in-built wireless adapters can start a wireless connection to a backbone wired LAN by simply coming within the range of a server's WLAN adapter or an access point. This method of communications can be extremely useful within large open areas, such as offices, a campus, and even warehouses. Temporary needs for LAN extension or connectivity failures can be handled with relative ease without rerunning wires or detailed cross-connects in the LAN wiring closets. Entire LANs can be automatically configured from a central site and made available to remote sites. All that the latter need to do is to turn on their computers equipped with a wireless LAN adapter and obtain instant access to the LAN. This over-the-air transmission mechanism is ideal when a company has a large number of remote sites, because the use of the technology can alleviate on-site visits, as well as the cost of supervising wiring or troubleshooting wiring problems. Although wireless solutions operate at lower speeds than their wired counterparts, the IEEE 802.11 standards provide data transfer rates up to 54 Mbps, which is comparable to many wired Ethernets.

13.2   MEDIA CONSIDERATIONS

Two types of media are standardized by the IEEE for wireless LANs:

1. Infrared (IR)
2. Radio frequency (RF)

13.2.1   IR Systems

One of the key advantages of IR is the fact that its use is unregulated. Like visible light, IR reflects off light-colored objects. This means that it is possible to use ceiling reflection to cover an entire room. However, similar to visible light, IR will not penetrate walls. This restriction in the transmission of IR can also be an advantage, as it offers a modicum of security against potential eavesdroppers.

A number of rooms each equipped with a separate IR system will not interfere with each other. This means we can create large LANs by coupling the different and independent room systems. IR is also relatively inexpensive and easy to deploy.

Because only amplitude modulation is used, wireless receivers do not need to monitor frequency and phase changes, as is required when an RF-based LAN transmission technique is used. The main disadvantage of IR is ambient radiation within rooms, resulting from sunlight or light bulbs. These are picked up as noise and can result in transmission throughput problems when a high level of optical noise occurs.

IR can be deployed using one of three techniques:

1. Directed beam
2. Omnidirectional
3. Diffused

13.2.1.1   Directed Beam IR

Directed beam IR is used for point-to-point communications links. The range for this approach depends both on the aim of the beam, as well as emitted power. Whereas directed beams can range up to a few kilometers, this capacity is needed only when connecting bridges or routers in a line of sight across different buildings.

13.2.1.2   Omnidirectional IR

In this situation, a single base system is needed with a line of sight to all stations on the LAN. Typically roof mounted, the base system serves as a repeater similar to a hub in 10BASE-T and 100BASE-T systems. The path from the central transmitter emanates to all transceivers within the area, whereas the latter all focus their beams onto the roof base station.

13.2.1.3 Diffused IR

In a diffused IR system, there are a number of roof-fitted base stations, one to a room. Each of these supports a number of stationary and mobile workstations. There is backbone wiring in the ceilings, whereby all base stations are connected to each other and to a central server that acts as a point of access to wired LANs and WANs.

Although IR systems are contained in most modern personal digital assistants (PDAs), laptops, notebooks, and even some cell phones such as the Nokia® 3660, the technology has only received limited success for use in LANs. Perhaps part of the problem resides in the movement of personnel in a modern office environment, where such movement can temporarily block line-of-sight IR transmission. In addition, typical office moves are not easily handled by IR systems, whereas in an RF environment the computer users only need to move their computer and wireless adapter to a new location within range of an access point to regain LAN connectivity. Thus, let us turn our attention to RF-based LANs.

13.2.2 RF LAN Networks

RF-based LANs use two leading technologies:

1. Spread spectrum
2. Narrow band

13.2.2.1 Spread Spectrum

This technique was originally invented for military and intelligence agencies. The basic idea is to spread out the data over a wider bandwidth to render interception and malicious corruption (jamming) more difficult.

Within spread spectrum, two subtechniques may be deployed:

1. Frequency hopping
2. Direct sequence

13.2.2.1.1 Frequency Hopping Spread Spectrum

Frequency hopping uses a pseudorandom sequence of radio frequencies with a translation (hopping) from frequency to frequency over time.

During the time when a specific (randomly generated) frequency is in use, the data may be modulated using an encoding technique such as

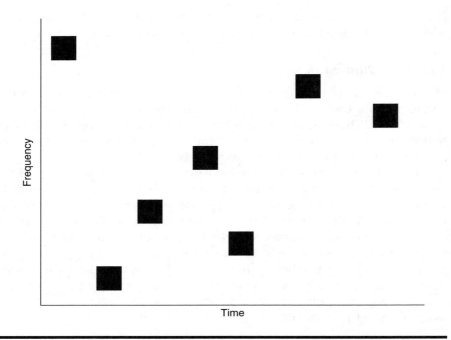

Figure 13.1 Frequency Hopping

frequency shift keying (FSK) or binary phase shift keying (BPSK). Figure 13.1 illustrates an example of the hopping of a signal across the frequency bands over time.

Under frequency hopping spread spectrum, as defined for use by the IEEE in 802.11 WLANs, a number of channels are allocated for the frequency hopping signals. The transmitter operates in one channel at a time for a fixed time interval. A random number generator functions as an index into the frequency set. Upon reception, the spread spectrum signal is demodulated using the same series of frequencies to reproduce the original data set.

13.2.2.1.2 Direct Sequence Spread Spectrum

Under direct sequence spread spectrum (DSSS) each data bit is multiplied by a spreading code. Each of the bits in the resulting codeword formed by the spreading code are modulated and transmitted. At the receiver, the spreading code bits are demodulated and compared to one another with a majority rule decision criteria applied to recovering the composition of the actual bit value. For example, if the bit to be transmitted is a binary 1 and the spreading code results in five bits being transmitted that are received as the values 11101, then because four of the five bits were set, the majority rule means that the bit received was assumed to have the value of binary 1.

One popular scheme for DSSS works as follows. Perform an exclusive OR (EOR) of the incoming signal with the bits in the spreading code (as series of 1s and 0s). This EOR operation results in an encoded transmission signal. At the receiver, a locally generated pseudorandom bit stream, identical to the one at the sender, is used to retrieve the original data from the direct sequence.

13.2.2.2 Spread Spectrum Configuration

Except for small organizations, spread spectrum wireless LANs use a multiple-cell configuration. Within a cell, the situation is either peer-to-peer or hublike with an access point (AP) functioning similar to a conventional wired hub.

In a centralized environment, the AP is connected to a backbone wired LAN, providing connectivity to both the wired station on the backbone and wireless clients. The AP can also centrally control access, in which case all stations may transmit to the AP and in turn receive solely from the AP.

When an AP detects a weak signal from a station, it can sense that that station has moved away to another AP's territory and then relinquish control to the next nearest AP.

In the second scenario (peer-to-peer), all stations may broadcast to all other stations by using an omnidirectional antenna, resulting in only the intended receiver accepting the data. Today, most wireless LANs operate in a centralized environment within two common frequency bands, referred to as industrial, scientific, and medical (ISM) bands. One band is in the 2.4 GHz frequency area, which includes microwaves and cordless telephones. The other band is in the 5 GHz frequency area, which currently has less interference in certain countries.

13.2.2.3 Narrowband RF

The name of this technique pertains to microwave RF with a relatively narrow bandwidth just enough to host the signal.

All narrowband microwave products have traditionally operated in licensed bands, with some recent forays into the ISM band. In a narrowband RF environment, neighboring cells use nonoverlapping frequency bands, which guarantees noninterfering communication. This is a level of QoS that license holders demand from their licensing authorities.

As observed, wireless LAN products are foraying into the ISM parts of the spectrum. At least one product works at 10 Mbps in the 6 GHz band, with a range of 50 m in semiclosed areas and 100 m in the open. Instead of using a fixed AP, a dynamically elected master is used, which during its tenure functions as the hub, until a more suitable master is elected. When source and destination are out of range, data moves from station to station until the source and destination get in range of each other.

The situation is akin to the analogy in the real world concerning individuals and post offices. When a person lives locally, the associated post office delivers and sends their mail. When someone moves to a new locality, they must deregister with the old post office and register with the new post office. When a mobile agent moves from its home network to a foreign network, it informs the home agent on the home network to that effect. Then all incoming messages intended for the mobile agent are forwarded by the (old) home agent to the new foreign agent.

13.3 TRANSMISSION ISSUES

Does a wireless LAN need licensing? Licensing authorities vary from country to country. In the United States, three RF bands have been set aside by the FCC for unlicensed usage:

1. 902 to 928 MHz (915 MHz band)
2. 2.4 to 2.4835 GHz (2.4 GHz band)
3. 5.725 to 5.825 GHz (5.8 GHz band)

Note that at around 900 MHz, one sees cordless phones, while microwaves operate in the 2.4 GHz band. And at the present time, there is little competition at the 5.8 GHz band. In general, the higher the frequency, the costlier the associated equipment. In addition, because higher frequencies attenuate more rapidly than lower frequencies, this law of physics means that APs operating in higher bands have a lesser area of coverage.

13.4 WLAN TOPOLOGY

In a wireless LAN environment the smallest artifice is the basic service set (BSS), consisting of a number of stations sharing a MAC protocol and competing for shared medium access. A BSS can be either isolated or connected to the backbone system via an AP. A BSS is often referred to as a "cell."

An extended service set (ESS) consists of two or more BSSs (cells) interconnected by a distribution system (DS), which is typically a wired backbone LAN. As previously noted, an AP is similar to a wired hub. Technically, the AP is a station that organizes all of the other stations in the WLAN around itself. It acts as a conduit for all these stations to communicate. In fact, if an AP is present there is no peer-to-peer communication.

Three types of station are defined:

1. No transition — a station of this type is either stationary or moves only within the range of other stations within a BSS.
2. BSS transition — defines a station movement to another BSS but within the same ESS. Delivery of data to that station requires recognizability of the addressing at the new location.
3. ESS transition — movement from a BSS in one ESS to a BSS in another ESS.

Stations may move, but maintenance of higher level connections supported by the IEEE 802.11 standard cannot be guaranteed and disruption of service can occur.

In a WLAN, the switch or hub connects APs together. This creates a wired backbone enabling WLAN roaming protocols to work.

Most APs accommodate connections to a switch or hub via an RJ-45 connector or a UTP Cat. 5 cable that can be up to 100 m long. One needs to carefully plan hub and switch installation to stay within this constraint. If distances are exceeded, one can use optical fiber to interconnect switches and place switches closer to APs.

The IEEE 802.11 standard defines a DS that provides interconnections between APs. The DS can be of any technology, such as Ethernet or Token Ring. At present, the majority of DSs are formed via the use of an Ethernet network.

When deploying a WLAN DS, some design considerations apply:

- Use hubs for smaller deployments — if there are only a couple of APs, a hub can serve the purpose. There is no need to invest extra funds for a switch.
- Switches for enterprise deployments — a larger WLAN with many APs may benefit from Ethernet switches. For very large networks, consider implementing a master switch interconnecting the client switches. Consider also interconnecting switches with optical fiber for greater range.
- Select the right data rate — in most cases, 10 Mbps Ethernet will suffice to interconnect 802.11b access points. For 802.11a type WLANs that operate at up to 54 Mbps and the more recent 802.11g standard that can operate at either a maximum of 11 Mbps or 54 Mbps, one will need a 100 Mbps Ethernet wired network.
- Create a separate IP domain for the WLAN — some network devices constantly send broadcast packets that freely propagate through Ethernet networks. APs will also forward these broadcast packets to all users on the WLAN. This can potentially flood and overwhelm

the WLAN, thus degrading the performance therein. One should consider separating the WLAN from the rest of the network using a router or a separate VLAN definition.

13.5 WIRELESS STANDARDS

Unlike in other 802.x scenarios, wherein different vendors competed with each other, 802.11 WiFi standards are in relatively good shape. IEEE alone sets worldwide baseline standards and there are only a handful of these.

A word on terminology used in this alphabet soup of standards is in order. WiFi or 802.11 is broken into two basic components:

1. Physical (PHY) layer — handles transmission and reception issues
2. MAC layer — governs media access

In essence, the PHY layer is the radio and the MAC layer is the software side of the receiving device (e.g., laptop, PDA, etc.).

The PHY standards are designated 802.11a, 802.11b, and 802.11g. They define data transfer rates and the frequency at which the data rides the airwaves.

Standard 802.11b came first, at 2.4 GHz and 11 Mbps, followed by 802.11a at 5 GHz and 11 Mbps. The more recent 802.11g standard defines operations on the same frequency as the 802.11b standard, but supports data rates up to 54 Mbps. To be backward compatible with 802.11b devices and networks, which are most commonly deployed, 802.11g works with this equipment.

Finally, one more, further-down-the-line backbone protocol being deliberated is the IEEE 802.11n, but at the moment the only thing known is that this protocol will support data transfer rates of up to 100 Mbps.

The original interest in standardizing a WLAN was geared to developing an ISM-based WLAN, using a token passing MAC protocol. This originally started as work in 802.4, but has since been superseded by the 802.11 series of standards. Again, this standard covers both the PHY and MAC layers. There is only one standardized MAC layer, which interacts with three different PHYs. Note that the wireless MAC, in addition to its usual functions, also performs such tasks such as fragmentation, packet retransmission, and handling acknowledgments. Two access methods are defined:

1. Distributed coordination function (DCF)
2. Point coordination function (PCF)

The MAC protocol standardized in 802.11 is sometimes called Distributed Foundation Wireless MAC (DFWMAC). This is basically a CSMA/CA mechanism. In this protocol, when a node receives a packet to be transmitted, it first listens to the network to ensure that no other node is simultaneously transmitting. If the channel is clear, it transmits the packet. Otherwise, it chooses a random back-off factor that determines the amount of time the node must wait until it is allowed to transmit its packet.

During periods when the channel is clear, the sending node decrements the back-off counter (unlike when the channel is busy, when the counter is not decremented). If the back-off factor reaches 0, the node transmits. Because the probability is small that two nodes will choose the same back-off factor, collisions are minimized.

Note that collision detection, as employed in Ethernet, cannot be used for RF transmissions because when a node is transmitting, it cannot hear any other node in the system that may be transmitting, as its own signal will eclipse any other signals arriving at the node.

Under the 802.11 standard, stations operate in one of two configurations:

1. Independent
2. Infrastructure

13.5.1 Independent Configuration

The stations communicate directly with each other, without any extra infrastructure. This is also appropriately called the ad hoc configuration. Figure 13.2 depicts this situation wherein each of the wireless stations communicates directly with its peers.

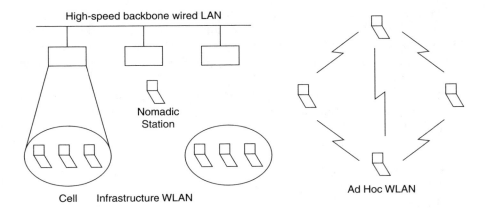

Figure 13.2 Infrastructure and Ad Hoc WLANs

13.5.2 Infrastructure Configuration

Stations communicate with APs that are a part of a DS. An AP serves the stations in the BSS.

Figure 13.2 illustrates this situation as well, as a cell with AP servicing cells (BSS) in relation to the high-speed backbone wired LAN.

The 802.11 standard defines a number of services that need to be provided by the WLAN, to offer functionality similar to that of a wired LAN.

- Authentication — used to identify stations to each other. Unlike a wired LAN, connectivity is achieved by simply tuning an antenna.
- Deauthentication — as the name indicates, this service deidentifies agents.
- Privacy — unintended target stations are not able to catch signals for the genuine receiver.
- Association — enables a link between a station and an AP. To identify and locate itself on a WLAN, a station must register with APs, which share that information with other APs.
- Disassociation — terminate an existing association.
- Reassociation — transfer an existing association to another AP.

13.6 WLAN DESIGN CONSIDERATIONS

Diffused IR technologies afford us a data rate of 1 to 4 Mbps. Regarding mobility, this method is either stationary or partly mobile and has a range of 50 to 200 ft. Detectability is negligible. Wavelengths used are of the order of 800 to 900 nm. The modulation technique is amplitude shift keying (ASK). Radiated power is marginal and the CSMA access method is used. No licensing is required.

Directed beam IR, on the other hand, has a data rate of 1 to 10 Mbps. It is stationary at the line of sight and supports a range of 80 ft. Again, detectability is negligible and the same 800 to 900 nm wavelength range is used. Modulation is also ASK and radiated power is marginal if any. The access method is CSMA and no licensing is required.

Frequency hopping spread spectrum gives us a data rate of 1 to 3 Mbps. Fully mobile, the range of this technique is 100 to 300 ft, with little detectability.

Frequencies involved are:

- 902 to 928 MHz
- 2.4 to 2.4835 GHz
- 5.725 to 5.85 GHz

The modulation technique is FSK and radiated power is less than 1 W. Access method is CSMA and no licensing is required.

DSSS affords us a data rate of 2 to 54 Mbps. Mobility criteria are either stationary or partly mobile and a range of 100 to 800 ft. is achieved. There is little detectability and the frequencies involved are:

- 902 to 928 MHz
- 2.4 to 2.4835 GHz
- 5.725 to 5.85 GHz

Modulation is via QPSK and radiated power less than 1 W. The chosen MAC protocol is CSMA and no licensing is required.

Finally, with narrowband RF, data rates of 10 to 20 Mbps are achieved. Either stationary or fully mobile options are available and a range of 40 to 130 ft is supported. There is some detectability and the frequencies involved are:

- 902 to 928 MHz
- 5.2 to 5.775 MHz
- 18.825 to 19.205 GHz

Modulation is via either FSK or QPSK and the radiated power is 25 mW. Access method is ALOHA or CSMA and licensing is required unless ISM.

13.7 WIRELESS LAN SWITCHING

Some of the traditional problems associated with WLAN deployment are aggravated cost, addressing network outages, and interface and user connectivity. All these required proficient RF experts. There was also no real way to obtain simple network information, such as the state of wireless APs, lists of users and devices, and network traffic statistics. These problems lead to a reexamination of WLAN technology, paving the way for wireless switches. WLAN switching ameliorates all these problems. The intent was to have a comprehensive and innovative solution that addressed all the following requirements:

- Transparent mobility support
- Visibility to corporate airspace
- Proactive security from rogue APs
- User authentication
- Traffic encryption
- 802.11 packet capture and monitoring
- AP management and automation
- Wireless or enterprise class of service
- Seamless interoperability with existing infrastructure
- Resilience and recovery

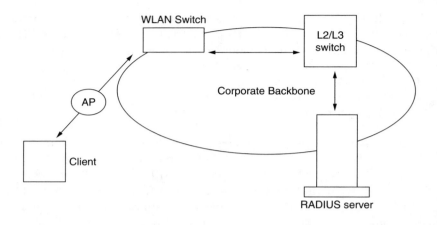

Figure 13.3   WLAN Switching Illustrated

In the WLAN switching approach, AP intelligence is integrated within a single WLAN switch. APs are simple, doing only transceiving and air monitoring functions. When connected to the WLAN switch directly or over a Layer 2/3 network, they essentially become extended access ports on the WLAN switch, passing on user traffic for processing while carrying out airspace control and visibility.

With the ability to manage AP power control and channel settings, WLAN switches also sense failed APs and direct nearby APs to adjust their power and channel settings to handle the information for the failed AP appropriately.

In the WLAN switching system, there are three layers:

1. Mobility management — this layer integrates mobile IP and DHCP with security management for switches. It provides security functions such as user authentication and mobile firewalls to maintain user identity, access control policy, and connectivity state across the wireless infrastructure.
2. Security management — this layer encrypts data and supports virtual private networks (VPNs) for all ports. In connection with air traffic management, this layer precludes airspace intrusion by rogue APs.
3. Air traffic management — this layer provides visibility into the enterprise airspace. It controls wireless bandwidth and manages traffic load to deliver to users class of service and performance management.

Figure 13.3 illustrates the following mode of operation within a LAN switch. The following occur therein:

1. Client sends 802.11 association request, which is automatically forwarded by the AP to the WLAN switch.
2. WLAN switch responds with association acknowledgment.
3. Client and WLAN switch start 801.x authentication conversation along with RADIUS server.
4. Encryption keys passed on to the WLAN switch and user device's own encryption keys, begins sending encrypted data.
5. WLAN switch decrypts data, processes packets, applies services, and forwards packets based on 802.11 MAC.

13.7.1 Additional Functions of WLAN Switches

Some of the additional features of WLAN switches include:

- Roaming the enterprise — this includes the ability of users to roam seamlessly between APs, WLAN switches, subnets, and VLANs without any change to existing infrastructure.
- Wireless intrusion prevention — if a new AP gets plugged into the wired LAN network, the WLAN switch can identify that device from both network and air space perspectives. Through air monitoring, the switch senses the beaconing of the new AP and registers it. If the AP is unauthorized, the switch immediately alerts the network administrator and acts forthwith on access policies.
- Automatic site survey — such technologies have largely been manual. WLAN switches allow network managers to carry out dynamic site surveys, from their desks, concerning:
  - Connection rates (e.g., 1 Mbps, 2 Mbps, 5.5 Mbps, 11 Mbps)
  - Channel plan (802.11a, 802.11b, or both)
  - Per AP user limit
  - Total number of users
  - Redundant overengineering

14

LOCAL AREA NETWORK INTERNETWORKING ISSUES

14.1 INTRODUCTION

One uses the term "internetwork" or simply "internet" to denote an arbitrary collection of networks interconnected in some fashion to provide host-to-host connectivity and deliver a service. For instance, an organization might have a number of sites, each implementing a LAN solution, and they might decide to interconnect these LANs using point-to-point links.

This term internet needs to be distinguished from the term Internet, which represents the global interconnection of many existing networks, including 802.3, 802.5, and even ATM. In certain circles, the preceding networks are termed "physical networks," whereas a collection of connected physical networks is termed as a "logical network." In this context, a collection of LANs connected by switches and bridges is still one network, whereas a collection of networks connected by routers is called an internet. Figure 14.1 illustrates a typical internet scenario. The key tool for managing internets is IP.

14.2 OVERVIEW OF INTERNETWORKING CONCEPTS

Network designers are faced with a daunting task when constructing an internetwork, because it is possible to use a mixture of four hardware devices:

1. Hubs (concentrators)
2. Bridges
3. Switches
4. Routers

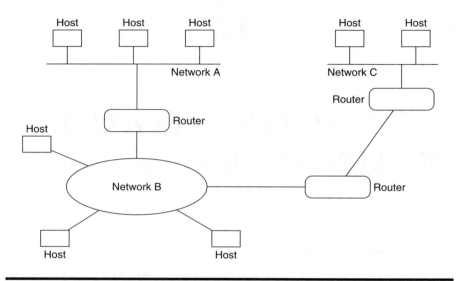

Figure 14.1   Internetworking Scenario

These were all discussed in detail in Chapter 8, but we quickly recall here their key properties, for the sake of continuity and completeness.

Hubs are used to link multiple users to a single physical unit, in turn connecting them to the network. They simply regenerate incoming signals out all ports, other than the port the data was received on, to all the attached stations.

Bridges serve to subdivide segments within the same network. They too function at Layer 2, independent of the network layer and other higher level protocols (Layer 3 and above).

Switches have more ports than bridges and can be considered to represent multiport bridges with added intelligence. If the number of ports is N, each operating at 10 Mbps, then the switch separates collision domains and provides an overall throughput of $10 \times N/2$ Mbps. Thus, while switches protect existing cabling infrastructure, they do increase performance and bandwidth.

Routers separate broadcast domains and connect disparate networks. Driven primarily by the IP protocol, routers make forwarding decisions based on IPv4 address formats rather than on link-layer MAC addresses.

The trend today is to move away from bridges and hubs and on to routers and switches when designing internets.

14.3   SWITCHING OVERVIEW

Switching data frames can occur via one of the following techniques:

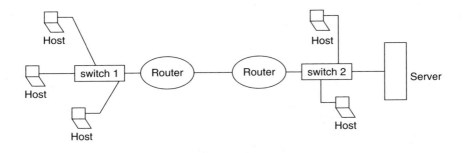

Figure 14.2   Routers and Switches

- Store and forward
- Cut-through
- Fragment-free
- Hybrid

These concepts were all introduced in Chapter 8 and will not be discussed again.

Maintaining switch operations denotes the build and maintenance of switching tables, route tables, and service tables.

Switching occurs at Layer 2 and routing at Layer 3. In other words, switches work based on the contents of 6-byte MAC addresses and routers use the 4-byte IPv4 or 16-byte IPv6 addresses.

Switches automatically build and maintain Layer 2 switching tables to track and learn MAC addresses. If a destination MAC address is not known, the switch broadcasts that frame out all ports other than the port the frame arrived on. By noting the source address of the frame and the port it arrived on, the switch updates its internal tables via a backward learning process. In comparison, routers are configured with the IP address of the networks attached to their ports and operate based on 4- or 16-byte IP addresses.

With increasingly bandwidth-hungry applications on the market, hubs in wiring closets are rapidly being replaced by LAN switches. There is also an increasing demand for intersubnet communications, which must flow through a router. In this connection, Figure 14.2 depicts a typical situation showing the relationship between routers and switches. Switches primarily move data within a local geographical area, such as a building. In comparison, routers provide long-distance and local interconnectivity.

Data flow from many hosts are passed serially through routers, which means that if there is significant traffic for a server accessed remotely via routers, there is the ineluctable possibility of congested bottlenecks at the routers.

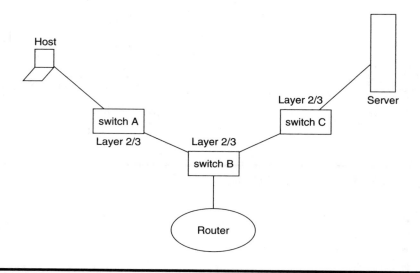

Figure 14.3   Flow of Intersubnet Traffic with Layer 3 Switches

To partly alleviate this problematic situation, Layer 3 capabilities are being added throughout many networks, typically within formerly Layer 2 switches. Figure 14.3 depicts such a scenario.

14.4   THE TIERED (LAYERED) APPROACH

Both ISO OSI and TCP/IP reference models are instances of a hierarchical approach to designing networks. Each layer is ascribed a set of functions or responsibilities that it provides as services to the layer above. Internetwork design uses a hierarchical tiered approach to help simplify the overall task. The advantage of a hierarchical design is modularity, which allows different elements in the tier to be independently constructed by different vendors and used mutually in an interoperable fashion This also facilitates management of change in the internet by containing the cost and complexity thereof.

Traditionally, hierarchical internet design uses three tiers:

1. Backbone (core) tier — optimal intersite communication
2. Distribution tier — policy-based connectivity
3. Local-access tier — user access to the network

The core tier provides high-speed packet switching without any time-consuming packet manipulation (e.g., filtering, error checks).

The distribution tier interfaces with the core and local access tiers. It manipulates data from the local-access tier and passes it on to the backbone.

Some of the functions of the distribution tier include:

- Address or area aggregation
- Department or work group access
- Broadcast (and multicast) domain definition
- VLAN routing
- Media translation
- Security

Some of the functions of the local-access tier include:

- Shared bandwidth
- Switched bandwidth
- MAC layer filtering
- Microsegmentation

14.5 EVALUATING BACKBONE CAPABILITIES

The evaluation of the backbone capability of a tiered network is extremely important, because it represents the primary data path. In this section, we will discuss the following:

- Path optimization
- Traffic priorities
- Load splitting
- Alternative paths
- Tunneling

14.5.1 Path Optimization

Recall that in computer networks, there are two types of protocols:

1. Route protocols
2. Routing protocols

The former have essentially to do with addressing techniques, whereas the latter pertain to trajectory selection from a fabric of paths available via routers and other networks.

Convergence occurs when there is a change in the network properties and all routers subsequently agree upon the optimal routes. This action takes place by means of neighbor greeting and autoconfiguration.

Routing protocols, also introduced in Chapter 8, come in two varieties:

1. Metric optimizing protocols
2. Policy-based protocols

Examples of the former are RIP and OSPF. An example of the latter is BGP. IGRP uses a hybrid metric based on bandwidth, load, and delay. Link State protocols like OSPF and IS-IS minimize the cost associated with the selected path.

14.5.2 Traffic Prioritizing

Whereas some networks can prioritize homogeneous internal traffic, routers prioritize heterogeneous flows. Such categorization is differentiated treatment, which ensures that critical data are given an edge over less important flows.

There are three types of category queuing:

1. Priority
2. Custom
3. WFQ

14.5.2.1 Priority Queuing

Traffic is categorized by a specific metric, such as protocol type. Typically, four output queues are used:

1. High
2. Medium
3. Normal
4. Low priority

Figure 14.4 illustrates an example of priority queuing. Note that UDP, which is typically represented by small segment lengths, such as DNS queries, is shown to receive high priority in this example. Most Layer 3 switches and routers permit the administrator to easily define data assigned to different queues.

14.5.2.2 Custom Queuing

Custom queuing provides more granularity than priority queuing, wherein multiple higher layer protocols are supported. Custom queuing reserves a portion of the bandwidth for a certain protocol, guaranteeing a predetermined bandwidth for it. Figure 14.5 illustrates an example of custom queuing.

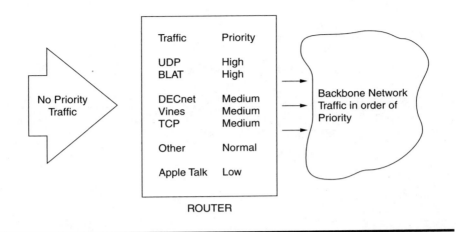

Figure 14.4   Priority Queuing

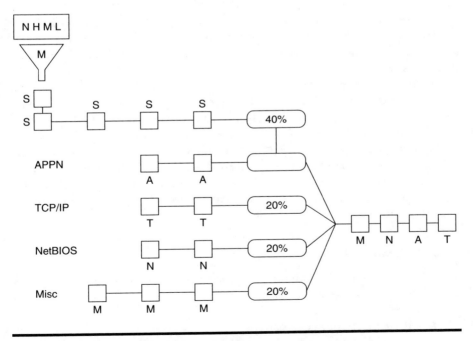

Figure 14.5   Custom Queuing

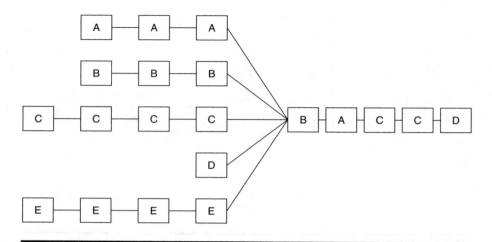

Figure 14.6   Weighted Fair Queuing

14.5.2.3   *Weighted Fair Queuing*

The WFQ method uses TDM to segment the available bandwidth among the several clients on the interface. By assigning weights, each client gets a weighted (for example, ToS) treatment based on a defined metric, such as arrival rates. Note that if all arrivals are assigned equal weights, low-volume traffic gets an edge over high-volume traffic. Figure 14.6 illustrates WFQ.

WFQ uses an algorithm to dynamically identify data streams at an interface and sort them into logical queues. Note that in certain cases, such as with SNA, one cannot distinguish between sessions. In DLSW+, SNA traffic is multiplexed over a single TCP session. In APPN they are multiplexed onto one LLC2 session. Because WFQ treats these sessions as a single conversation, the algorithm does not lend itself to SNA.

In priority queuing and custom queuing, access lists need to be preinstalled. However, this is not the case with WFQ, which sorts among specific traffic streams in real-time.

14.5.3   Load Splitting

This is exactly what the name implies, load balancing over different paths. Load splitting can be done with:

- IP (using equal cost paths)
- (E)IGRP (with unequal cost alternatives)

Up to four paths may be used for one destination network. Load splitting of bridged traffic over serial lines is also possible.

14.5.4 Alternative Paths

The necessity here is to provide for complementary paths to a destination, in case of link failures on active networks. End-to-end reliability is achieved only when there is redundancy throughout the network. Because redundancy is so expensive, most providers support redundancy on segments carrying mission-critical data.

Routers are the key to reliable internetworking. However, merely making hardware at the nodes more available does not make the internet reliable.

Instead, it is necessary to have redundant links as well. Unless all backbone routers are fault tolerant, it is necessary also to ensure that redundant links should terminate at different routers. Thus, a fully fault tolerant router situation is not only prohibitively expensive, it does not address the link reliability issue. We will return to reliability options later.

14.5.5 Encapsulation (Tunneling)

Encapsulation or tunneling is a simple operation, which takes packets or frames from one network and hides them within frames from another protocol.

14.6 DISTRIBUTION SERVICES

We include a discussion of the following functionalities:

- Backbone bandwidth management
- Area and service filtering
- Policy-based distribution
- Gateway service
- Interprotocol route redistribution
- Media translation

14.6.1 Backbone Bandwidth Management

To optimize use of the backbone, routers are able to offer features such as:

- Priority queuing
- Routing protocol metrics
- Termination of local sessions

Metrics on queues, overflow mechanisms, and routing protocol are all adjustable to gain more control over forwarding packets through the internet. If a local session terminates, a router can proxy for it instead of passing through all session control to the multiprotocol backbone.

14.6.2 Area and Service Filtering

This functionality is achieved by the use of access lists, which control the movement of data based on, among other things, network addresses. Service protocols are applicable to specific protocols.

14.6.3 Policy-Based Distribution

A policy in our context is a set of rules governing end-to-end traffic to a backbone network. For example:

- A LAN department may send traffic to the backbone using three different protocols, whereas it may wish to expedite one specific protocol through the backbone as it contains mission-critical data.
- Another department may wish to exclude all but remote login and e-mail from entering its LAN.

These are departmental policies, and organizational policies can exist as well. For example, an organization might decree that no Web-based e-mail should enter or leave its intranet.

Different policies may require different internetworking technologies, which may all need to be integrated and coexist harmoniously.

One possible way to implement policies is via SAPs (service access points). This situation is depicted in Figure 14.7.

In Figure 14.7, SAPs from the NetWare® servers advertise services to clients. Depending on whether services are provided locally or remotely, SAP filters prevent SAP traffic from leaving the router interface.

14.6.4 Interprotocol Route Redistribution

The section above on gateway services related to two end nodes using different route protocols to be able to communicate. Meanwhile, routers can interchange different routing protocols (RIP, OSPF, IGRP, etc.), which exchange routing information at the router. Static routing information can also be redistributed.

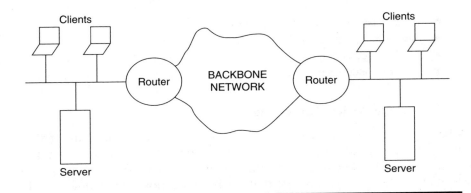

Figure 14.7 Policy-Based Distribution: SAP Filtering

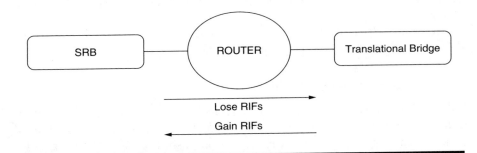

Figure 14.8 SR/TL Bridging Topology

14.6.5 Media Translation

These are techniques that translate frames from one network system to another. If there are attributes in the one system with no counterpart in the other, we have a problem on our hands. Different vendors will make different decisions as to how to manage this situation. For example, when a direct bridging is sought between, for example, Token Ring and Ethernet, one uses either SR/TL or SRT bridging.

SRT allows the router to use both SR bridges and a transparent bridging algorithm. There is a standard way to convert between SR and translational bridges, as illustrated by Figure 14.8.

14.7 LOCAL ACCESS SERVICES

Topics we consider here include:

■ Value-added addressing
■ Network segmentation

- Broadcast or multicast capability
- Naming, proxy, and local cache
- Media access security
- Router discovery

14.7.1 Value-Added Addressing

When different addressing schemes exist for LANs, such as IP and NetWare, they interoperate less than perfectly over multisegmented LANs/WANs.

Interprotocol specific helper addressing is a method to move specific datagrams through a network that such traffic normally would not be allowed to transit. For example, a client may search for a server and then broadcast a message that must transit many routers. Normally such frames would be dropped, but helper addresses allow such messages to go past routers.

Multiple helper addresses are supported on each router interface to allow forwarding to remote destinations.

14.7.2 Network Segmentation

This is an instance of the usage of local access routers to implement local policies and thus limit unnecessary traffic by segmenting traffic within component segments. One way to accomplish this is by strategically positioning routers and building in specific segmentation policies.

For example, a large LAN might be subdivided into segments, such that traffic on a segment might be limited to:

- Local broadcasts
- Unicast intrasegment traffic
- Traffic for another specific router

Careful distribution of hosts and clients leads to reduced congestion in the network.

14.7.3 Broadcast or Multicast Capabilities

Routers intrinsically drop broadcast messages. But these are quite commonplace and need to be curbed to reduce traffic to a manageable level and reduce broadcast storms. Again, helper addresses help aid multicasts or broadcasts.

To fully support IP multicast, the IGMP (Internet Group Management Protocol) must be deployed on hosts. IGMP enables hosts to dynamically report their multicast group memberships to a multicast router.

Multicast routers send IGMP queries to their attached LANs and stations respond with their membership information. The multicast router attached to the LAN then takes responsibility for sending multicast datagrams from one attached network to all other networks with multicast membership. If an IGMP query brings no response, that group is deemed to have no members. No further messages are sent to that group in the future.

14.7.4   Naming, Proxy, and Local Cache Capabilities

These three router capabilities reduce traffic and enable efficient internet operation. They include:

1. Naming service support
2. Proxies
3. Local caching of network information

Naming is a well-known mechanism used to resolve names to addresses. Common examples of addressing schemes include:

- IP Domain Naming Service (DNS)
- Network basic input output system (NetBIOS)
- IPX

A router can proxy a name server. For instance, a list of NetBIOS addresses can be maintained, avoiding the overhead of transmitting client/server broadcast in an SR bridge environment.

In that case, the router does the following:

- Only one (duplicate) query frame is allowed per time period configured.
- A cache of NetBIOS server addresses with client names (and MAC addresses) is maintained, limiting broadcasts across the network.

14.7.5   Media Access Security

This serves to:

- Keep local traffic from inappropriately accessing the backbone
- Keep backbone traffic from inappropriately entering the LAN

Both problems need packet filtering to be alleviated. Packet filtering reduces traffic levels to improve performance. Also, as its name implies, this function improves security and reduces congestion. The most popular filtering mechanism is the access list approach.

14.7.6 Router Discovery

As its name implies, this service is a process of finding routers: ES-IS. ES is ISO terminology for host stations and IS pertains to routers. Limited to exchanges between hosts and routers, Hello messages are sent by ESs to all routers on the subnet and in reverse. Both carry subnet and Layer 3 addresses of their generating systems.

14.7.7 Internet Control Message Protocol Router Discovery

RFC 1256 outlines a process for ICMP. There is no single, standardized protocol for this mechanism.

14.7.8 Proxy ARP

A proxy-ARP-enabled router responds on behalf of all hosts that it has a connection with. This allows hosts to assume that all other hosts are on the network.

14.7.9 Routing Information Protocol

RIP is commonly available on hosts and is used to find the most suited router given an address.

14.8 CONSTRUCTING INTERNETS BY DESIGN

We start with backbone considerations:

- Multiprotocol routing backbone
- Uniprotocol backbone

When several Layer 3 network protocols are routed through a common backbone, without encapsulation, the situation is that of a multiprotocol routing backbone. Two strategies are available:

1. Integrated routing — uses one preferred routing protocol that determines the metric-minimizing path
2. Ships in the night — uses a different routing protocol for each route protocol

All routers support one specific routing protocol per specific route protocol. Encapsulate all other routing protocols within the preferred supported routing protocol.

14.9 USING SWITCHES (REVISITED)

Vendors and implementations are moving away from hubs and bridges to switches and routers. All switches operate at Layer 2 and have the following benefits:

- Superior segmentation
- Increased aggregate forwarding
- Increased backbone throughput

LAN switches address end host requirements for greater bandwidth. Rather than deploying hubs, by using switches, designers can increase performance and better exploit existing media. Also, previously unavailable functionality such as VLANs may become available when the functionality is incorporated into a switch. In addition, by delivering links to interconnect existing, shared hubs in wiring closets, and to server farms, a scalable bandwidth becomes available.

14.9.1 Comparison of Switches and Routers

To conclude our discussion of internetworking, we will summarize the major differences between switches and routers. Key features of switches include:

- High bandwidth
- High performance
- Low cost
- Easy configuration

Key features of routers include:

- Broadcast firewalling
- Hierarchical addressing
- Inter-LAN communication
- Quick convergence
- Policy routing
- QoS routing
- Security

- Redundancy
- Load splitting
- Traffic flow management

Note that as switch technology gains momentum, switches of the future will address all these router functionalities. Although routers currently have more features than switches, a new series of switches that include built-in routing capability deserve the consideration of network designers and analysts when constructing or revising a network.

INDEX

A

Access controller sizing, 173, 178
Access lines, 44, 46
Access points, 287
Accunet T45, 61
    bandwidth, 61
    break-even point, 61
    costs, 61
    data rates, 61
ACK/NAK, 14
Activity maps, 85
Actual data rates, 73
Adaptive packets, 21
Additional terminal devices, 33
Address Resolution Protocol, 98
Ambient radiation
Amplitude modulation, 288
Analog circuits, 3
Analog technology, 3
Apple Talk, 202
Applicability of traffic measurements, 152
Application of equipment sizing process, 184
ARP, see Address Resolution Protocol
Arrival rates, 179
ASIC, 199
Asynchronous channels, 70
Asynchronous terminal devices, 70
Asynchronous Transfer Mode, 4
AT&T
    ASDS (Accunet Sprectrum of Digital Services), 52, 57
    800 READYLINE service, 45
    WATS Service Area 1, 45
AT&T divestiture, 29, 38
    facilities, 32
ATM, see Asynchronous Transfer Mode
Autobaud, 96
Automation of MST, 135

Automation of the location process, 118
Availability levels 259–261
    increase in, 260
Average traffic intensity, 153

B

Backbone, 191
Backbone capability evaluation, 305
    alternative paths, 309
    load splitting, 308
    path optimization, 305
    traffic priorities, 306
    tunneling (encapsulation) 309
Backbone common circuit, 3
Badge reader system, 11
Bandpass multiplexing, 89
Bandwidth, 190
Bandwidth hungry applications, 303
Baseline service, 30
Basic connection matrix, 113
Basic Language program for modified MST, 136–139
Basic Language traffic analysis program, 161
Basic NODE analysis program, 122
Batch systems, 9
Batch transmission, 9
BH traffic intensity, 174
Binary Synchronous Communications Protocol, 13, 17
BISYNC, see Binary Synchronous Communications Protocol
Block, 10
Block check character, 14, 16
Block Size, 20, 21, 22
Bottom up approach, 4
Branches, 110
Break even point, 61
Bridge port and address table, 193
Bridged topologies, 192, 195

317

Bridges, 4, 191, 301–302
Bridging a network, 268–269
British Telecom KiloStream, 52
Broadband signaling methods
    amplitude shift keying, 211
    frequency shift keying, 211
    phase shift keying, 211
Brouters, 4, 205
Buffer control, 87
Buffer memory considerations, 264
Buffer/flow control, 93
Busy hour, 150, 155, 174, 185

C

Cable modems, 68
Calling population, 155
Call-seconds, 151
Carrier extension, 258
Cat(egory) 3–5, 231
Channel adapters, 72
Channel service units, 12
Circuit operating rate, 1
Clear-to-send (CTS), 16
Coax adapters, 232
Communication servers, 207
Concentrators, 64
Congestion control, 277
Connection matrix, 133
Constraints, 9
Control signal passing, 74
Convergence (times), 281
Cost of cabling, 35-36
Costing assumptions, 35
Cost-performance relationships, 58
Criteria for evaluating routing protocols, 279
CSU, see Channel service units
Cumulative delay, 14
Cut-through bridges/switches, 192, 303
Cycles, 113

D

Data blocks, 25
Data center, 10
Data concentration equipment/devices, 1, 3,
    41, 109
Data flow, 100
Dataphone digital service (DDS), 54
    costs of, 54

evolution, 54
    obsolescence, 54
    v/s Accunet T1.5, 57
Data service units (DSU), 12, 52, 73, 111
    DSU speeds, 52
Data source support, 89
Day rates, 31
Dead time, 14
Decision model, 149
DEC-Net, 202
Dedicated lines, 27, 28
Delay time calculations, for
    Ethernet, 285
    switches, 286
Delay time, 12
Delays, 5
Demultiplexing, 103
Design constraints, 4, 5
Design issues, 274
    addressing, 274–275
    datagram lifetime, 225
    fragmentation/reassembly, 275–276
    routing, 275
Device differences, 104
Dial-in lines, 3, 20, 31, 37, 144
Dial-up services, 33
Digital circuits, 3
Digital leased lines, 34
Direct connect lines, 28
Distributed computing, 63
Distributed terminals, 116
Distribution services, 309
    area/service filtering, 310
    backbone bandwidth management, 309
    inter-protocol route redistribution, 310
    media translation, 311
    policy-based distribution, 310
DSL, 68
DSU, see Data service units
Dual-port adapters, 145

E

Echoplexing, 97
Economic comparisons/analyses, 27, 36, 38
Economical routes, 127
Encapsulation, 194
Encoding
    baseband (digital), 211
    broadband (analog), 211

Encoding schemes,
   Differential Manchester, 249
   8B/10B, 251
   4B/5B NRZ-I, 249
   Manchester, 249
   MLT-3, 250
   NRZ-I, 249
   NRZ-L, 248
Equipment location techniques, 115
Equipment sizing 3, 143
Erlang B & C formulas, 144
Erlang traffic formulas, 156
Erlangs, 157
Error control, 14
Error rates, 20, 21
Estimation of response times, 141
Ethernet family, 4, 229, 234, 236, 238, 241
   Fast Ethernet, 236–237
   Gigabit Ethernet, 238–241
   10 Mbps Ethernet, 234
   10 Gigabit Ethernet, 241–243
Ethernet frame composition, 254–257
Ethernet performance details, 266
   actual operating rates, 268
   GE considerations, 267
   network frame rate, 266–267
Evening rates, 31
Expedited forwarding, 65
Experimental modeling, 144

F

Facility, 28
Fan-out, 132
Fast Ethernet (100 Mbps), 236–238
   backbone operation, 237–238
   frame operations, 253–259
   frames, 257
   switch segmentation, 238
FDM, 64, 66–68
Fiber channel, 226
   fabric, 228
   five layers of, 227
   F-Port, 228
   media, 228
   NL-Port, 228
   N-Port, 228
   routing, 230
   topologies, 229
Fiber optic technology, 233
   fiber adapters, 234

fiber hubs, 234
optical transceiver, 233
File servers, 4, 206–207
5-4-3 rule, 5, 232
5-level Baudot, 72
Fly back delay, 96
Foreign exchange (FX), 27, 48
   call origination, 48
   call reception, 48
   cost per minute, 50
   economics, 49
   FX line, 48
   local call, 48
   monthly recurring cost, 50
   utilization, 49
Formula comparison and utilization, 182
Four-position rotary, 160
Fractional T1 (FT1), 30
   costs associated, 60
Front-end processing techniques, 142
Front-end substitution, 79
Full duplex model, 13, 24
Future tariffs, 27

G

Gateways, 4, 65
   simple gateway function, 65
Gigabit Ethernet
   carrier extension, 240, 258
   frame composition, 257–259
   frame overhead, 259
   GMII, 238–240
   packet bursting, 240–241, 258–259
   1000 BASE-CX, 238
   1000 BASE-LX, 238
   1000 BASE-SX, 238
   1000 BASE-T, 238
Graph theory, 3, 109–115

H

Half-duplex models, 13, 15
HDLC, see High-Level Data Link Control
   protocol
High Speed Study Group (HSSG), 241
High-Level Data Link Control protocol
   (HDLC), 13
High-speed modems, 38, 41
Holding times, 153

Hub-bypass multiplexing, 78
Hubs, 4, 64, 190, 302
Hybrid bridges, 192

I

IEC, 30
IEEE 802.11x, 5, 287, 294
   independent configurations, 295
   infrastructure configurations, 296
Individual port traffic statistics, 176
Information transfer efficiency, 13
Information transfer rate, 11–13
Information transfer ratio (ITR), 17, 19, 18,
    21, 23, 24
Infra-Red (IR) systems, 288
   diffused IR, 289
   directed beam IR, 288
   omnidirectional IR, 288–289
Intelligent hubs, 190
Inter LATA, 29
Interactive transmissions, 9
   response times, 10
Interexchange carriers, 28, 29
Interleaving, 71
   character-by-character, 71
   teletype model, 33, 71
Intelligent hubs, 190
Internal delay time, 12, 14
Internet (large 'I'), 2, 301
Internet (small 'i'), 2, 301
Internets (small 'i'), by design, 314
   backbone considerations, 314
   integrated routing, 314
   ships in the night, 314
Internetworking, 5, 301
Interoffice channel charges, 42
Intra LATA, 29
Inverse multiplexing, 80
Ipv4, 102, 202
   addressing, 278–279
   header fields, 271
IS-IS, 202
ISO, 13
ISPs, 143
ITDMs, 91
   statistics, 91
   system reports, 92

K

Kruskal's algorithm, 195

L

LAN access controllers (sizing of), 143
LAN architecture evolution, 223–225
   design evolution, 224
   layers 1-3, 224
LAN devices, 189
LAN ethernet design, 243–246
   campuswide VLANs with multilayer
     switching, 245–246
   router on a stick, 244
   wire hub and router model, 244
LAN performance, 253
LAN switches, 4, 197
   cut-through, 198
   fragment free, 199
   hybrid, 199
   store and forward, 198
LAN topologies
   bus, 210, 213–214, 218
   loop, 209, 213, 218
   star, 210, 214–216, 219
   tree, 214, 218
Large capacity routers, 41
LATA access tariffs, 30
LATAs, see Local access and transport areas
Latency, 5, 247
Layer 2 (Data Link Layer), 194
Leased lines, 27, 28
Level of utilization, 11
Line occupancy, 94
Line sharing devices, 3, 34
Line turnarounds, 23, 25
Local access and transport areas, 29
Local area services, 311
   broadcast/multicast capacity, 312
   media access security, 313
   naming, proxy, local cache capability,
     313
   network segmentation, 312
   router discovery, 314
   value-added addressing, 311
Local exchange carriers (LEC), 28-30
Lost traffic computations, 159

M

MAC addressing, 98
MAC protocols, 220
    and Xerox, 221
    BIU, 221
    CSMA/CA, 222
    CSMA/CD, 220–221
    DPMA (100 VG-AnyLAN), 222–223
    suitability, 221
    switched connection-oriented, 222
    token passing, 222
Maximum data rates, 74
MCI, 29
Microcom Network Protocol, 20, 23
Mileage bands, 121
Minicomputers, 64
Minimum spanning tree, 128
MNP, see Micocom Network Protocol
Models, 9
Modem, 12, 34, 37, 63, 102–103
Modem sharing units, 106
Modified MST, 132
Molina (Poisson) traffic formula, 181
MST technique and algorithm, 128
Multidrop line, 1, 3
Multimedia service, 284
    H.323 recommendations, 284–285
    PBX components, 284
Multiplexed leased line, 37
Multiplexer buffer occupancy information,
    95
Multiplexer performance, 94
Multiplexers, 3, 34, 35, 39, 40, 63, 64
    statistical/intelligent, 64, 83–87
    traditional, 64
Multiplexing economics, 81
Multiplexing intervals, 70
Multiport bridges, 192
Multirate schedules, 39

N

Network adapters, 287
Network configuration, 36
Network constraints, 140
Network interface card, 11, 206
Network management, 5
Network subdivisions, 112
Network topology, 109
Network traffic estimation, 261

NIC, see Network interface card
Night rates, 31
Node location problem, 120
Nodes, 110
NTU, 52
NVRAM, 99

O

Open Shortest Path First, see OSPF
Optimal block size, 20
Optimal level of service, 150
Optimal routing path, 21
OSPF, 202
Other delay factors, 12
Outlay, 104

P

Packet bursting, 258–259
Packet switching, 100
Parity bits, 75
Partial full-duplex, 23
Performance considerations, 3, 4, 5, 53, 127,
    216, 231
Phantom channel, 64
Points of presence, see POPs
Point-to-point system, 76
Poisson formula, 177
Polling (round robin), 140
POPs, 29, 30, 42
Port switches, 192
Ports and connectors, 98
Prim algorithm, 128, 129–132
Print line, 10
Priority, 140
Private lines, 28
PRO WATS, 44
Processing and acknowledgment times, 16,
    25
Processing delay time, 12
Productivity, 147
Programmed ROMs, 72
Propagation delay time, 14, 25
Protocol architecture, 273
Protocol efficiency, 21
Protocol overhead, 75
PSTN calls, 33
PSTN facilities, 41
PSTN rate table, 31

PSTN system life, 39
PSTN usage, 35
Pythagorean theorem, 124

Q

QoS, 65
Quality of service, see QoS
Query, 10
Queuing theory, 263

R

Rack mounted LAN access controllers, 145
RAS, 207
Record, 10
Reliability, 53
Remote Access Servers, see RAS
Remote bridges, 197
Remote device, 3
Repeaters, 4, 190
Request to send (RTS), 16
Response times, 9, 10, 93
    Data rate, 23
RF LAN networks, 289
    narrowband, 291–292
    spread spectrum, 289
        configuration, 291
        direct sequence, 290–291
        frequency hopping, 289–290
Ring indicator signals, 74
RIP, 202
Route dimensioning parameters, 155
Route protocols
    vector distance, 204, 280–281, 282
    link state, 205, 281–283
Router behavior, 201
    addressing scheme, 201
    next hop, 201
    route protocols, 202
    routing protocols, 202
Router configuration information, 100
Router software modules, 100
Router types
    multiprotocol (intelligent) 203
    nonroutable protocol support, 203
    protocol independent, 203
    uniprotocol, 203
Routers, 3, 4, 34, 35, 39, 40, 63, 65, 97–102,
    200–205, 302

Routes, 112
Routing analysis, 127
Routing Information Protocol, see RIP
Routing revisited, 277
RS-232 interfaces, 74
RTMP, 202

S

Scalability, 247
Scientific approach (to sizing), 146
Secondary channels, 23
Segment switches, 192
Segments, 189
Series multipoint multiplexing, 77–78
Service ratio, 87
Sharing units and constraints, 105
Simple gateway function, 65
Sizing, 37, 143–144
Sizing methods, 144–146
Software default settings, 23
SONET, 52
Split horizon (with poison reverse), 281
Sprint, 29, 38
SR bridges, 191, 196
SR/TL bridges, 192, 196
SRT bridges, 191, 196
Stackable hubs, 190
Start/stop bits, 75
Station message processing time, 14
Stations, 189
Statistical frame construction, 84
STDMs, 83
    activity maps, 85
    buffer control, 87
    data source support, 89
    frame construction, 84
    selection features, 96
    service ratio, 87
    STDM statistics, 91
    switching and port contention, 90
    system reports, 92
Stop and wait ARQ, 14, 17
Storage requirements determination, a 9-
        step approach, 265–266
Store and forward bridges/switches, 303
Structured cabling system, 219
Subdividing switches
    port switches, 200
    segment switches, 200
Subscriber population, 155

Switch design
  bus architecture, 199
  matrix, 199
  shared memory, 199
Switched lines, 27
Switched network cost table, 121
Switched network utilization, 120
Switches and routers, compared, 315
Switches revisited, 246, 302
Switching packets, 97
Switching speed metric (pps), 101
Synchronous optical network, see SONET
System life, 35, 47, 50

T

TCP, 283
TDM, 64, 66, 68–70, 103
TDMA, 199
Telephone terminology relationships, 146
Telephone traffic formulas, 144
  layers and architecture, 242–243
  objectives, 241–242
10 Mbps LANs, 234–236
Terminal device, 34
3 F's, flooding, filtering and forwarding, 193,
      217
  backbone (core) layer, 304
  distribution layer, 304–305
  local access layer, 305
Tiered switches, 4
Time division multiplexing, see TDM
Time subdivision, 70
  applications, 76
  constraints, 72
Token Ring, 4
Top down approach, 4
Total delay time, 23
Total route distance, 126
Total-service circuit, 30
Traffic capacity planning, 165
  QuickBasic program, 167
Traffic carried, 159
Traffic dimensioning formulas, 156
Traffic estimation, 253
Traffic intensity, 154
Traffic lost, 159, 176
Traffic measurements, 150
Traffic tables, 171
Transaction, 10
Transfer rate, 12, 18

Transmission control structure efficiency, 16
Transmission errors, 33
Transmission media, 210
  coaxial cable, 231–232
  optical fiber, 232–233
  UTP/STP, 211, 230–231
  wireless, 233
Transmission rate of information in bits, see
      TRIB
Transmission throughput problems, 288
Transmission time, 23, 40
Transport circuit, 30
Trees, 113, 134
TRIB, 12
Trunks and dimensioning, 147–149

U

UART, 70
UDP, 283
Universal Asynchronous
      Receiver/Transmitter, see UART
U.S. Commerce Department, 29
Usage of LAN switches
  backbone, 200
  network redistribution, 200
  server segmentation, 200

V

Verizon, 38
Vertical/horizontal (VH) coordinate grid
      system, 123
VH boundaries, 124
Voice over IP (VoIP), 65
Voice-band truck, 44

W

WAN transmission requirements, 27
WATS, 27, 43, 144
  intrastate WATS, 45
  inward WATS, 43, 45, 46
  outward WATS, 43
  WATS and PSTN, 45
  WATS plans, 43
Weighted connection matrix, 117
Wide area telecommunication service, see
      WATS
Wide band facilities, 80

Wireless access methods, 294
  distributed coordination function (DCF),
    294
  point coordination function (PCF), 294
Wireless LAN adapter, 287
Wireless switches, 5
Wireless transmission issues, 292
WLAN switching, 297–299
  air traffic management, 298
  automatic site survey, 299
  mobility management, 298
  roaming the enterprise, 299
  security management, 298
  wireless intrusion prevention, 299

WLAN topologies, 292
  access points, 292
  basic service set, 292
  design issues, 293, 296–297
  distribution system, 292
  extended service set, 292
WLANs, 287

X

Xerox network system, XNS, 202